THE PROFIT OF THE EARTH

The Profit of the Earth

THE GLOBAL SEEDS OF AMERICAN AGRICULTURE

COURTNEY FULLILOVE

THE UNIVERSITY OF CHICAGO PRESS

Chicago and London

The University of Chicago Press, Chicago 60637
The University of Chicago Press, Ltd., London

Published 2017

Printed in the United States of America

26 25 24 23 22 21 20 19 18 17   1 2 3 4 5

ISBN-13: 978-0-226-45486-3 (cloth)
ISBN-13: 978-0-226-45505-1 (e-book)
DOI: 10.7208/chicago/9780226455051.001.0001

Library of Congress Cataloging-in-Publication Data
Names: Fullilove, Courtney, author.
Title: The profit of the earth : the global seeds of American agriculture / Courtney Fullilove.
Description: Chicago : The University of Chicago Press, 2017. | Includes bibliographical references and index.
Identifiers: LCCN 2016041539 | ISBN 9780226454863 (cloth : alk. paper) | ISBN 9780226455051 (e-book)
Subjects: LCSH: Seeds—United States—History—19th century. | Seeds—Harvesting—United States—History—19th century. | Plant introduction—United States—History—19th century. | Seed industry and trade—United States—History—19th century. | Wheat—Breeding—United States—History—19th century. | Plant diversity conservation—United States—History—19th century.
Classification: LCC SB187.U6 F85 2017 | DDC 631.5/21—dc23
LC record available at https://lccn.loc.gov/2016041539

♾ This paper meets the requirements of ANSI/NISO Z39.48-1992 (Permanence of Paper).

For how hard it is
to understand the landscape
as you pass in a train
from here to there
and mutely it
watches you vanish.
— W. G. Sebald

Moreover the profit of the earth is for all:
the king himself is served by the field.
— Ecclesiastes 5:9

CONTENTS

PROLOGUE : **IN THE FIELD**

I never liked the adage about starting at the beginning. History defies beginnings and ends, and in any case, margins are much more interesting points of departure: here, the border of a wheat field on the edge of a remote archaeological site near Sisian, Armenia. Zorats Karer (Army Stones) is a mountain plateau covered with 223 basalt stones planted upright, many with small holes bored in the upward ends. Archaeologists have speculated about astrological and religious purposes for the site. Ringed by petroglyphs in the surrounding mountains, the margins of the crop bear inscriptions from every stage of human history: from 200,000 years ago with the first humans in the middle Paleolithic period, to the domestication of plants in the Neolithic, to the fabrication of these monuments perhaps 10,000 years afterward in the Middle Bronze Age. These varied attempts to periodize human history link the passage of time to tool use, including agricultural implements. The implied narrative of human development from hunting and gathering to settled agriculture hints at the dense meanings that accrue to the plants in the field. Food plants are not simply products of nature, but complex by-products of human interactions with their environments, sustaining millennia of civilization.

This book traces how seeds and knowledge about them were acquired, melded, and trafficked within the context of the burgeoning, nationally and internationally oriented economy of the nineteenth-century United States, but it rests on a series of more basic insights about human history: (1) Cultivated seeds are not products of nature but deep-time technologies, domesticated some 10,000–12,000 years ago and improved by successive generations of farmers. (2) As proxies for multigenerational agricultural labor, seeds simultaneously contain and obscure the complex social relations required for their production. (3) Although agriculture itself is a radical diminishment of biodiversity, we can distinguish between different kinds of agricultural practices and their relative impacts on the environment. (4) Innovation, inevitably, is a process of destruction and loss, and this volatility partially accounts for the recursive qualities of discovery and invention.

Figure 0.1. ***A*****, Zorats Karer, near Sisian, Armenia.** ***B*****, Wheat field of traditional (landrace) and modern (cultivar) varieties adjacent to Zorats Karer, Armenia. Photos by Courtney Fullilove.**

In exploring these propositions, I became a historian, and then, to my surprise, I became a collector. Since 2010, I have accompanied plant genetic resource specialists from the International Center for Agricultural Research in the Dry Areas (Aleppo, Syria, and Rabat, Morocco), the Vavilov Research Institute (Saint Petersburg, Russia), and AgResearch (Palmerston North, New Zealand), collecting locally adapted varieties of cereal and legume crops (landraces) and their wild relatives for nationally and internationally managed gene banks.[1] Together we have traveled to mountainous areas of the North and South Caucasus and Central Asia, visiting the post-Soviet republics Tajikistan, Kazakhstan, Armenia, and Georgia, as well as the semiautonomous Russian republics of Karachay-Cherkessia, Kabardino-Balkaria, and Adygea. The plants we gather are screened for resistance to biotic and abiotic stressors,

with an eye toward breeding varieties resistant to pests, disease, and climatic variation. Lodged in nationally and internationally managed gene banks at the sponsoring institutes, the Svalbard Global Seed Vault in Norway, and the countries of collection, the seeds we collect provide material for research, and they represent part of an ambitious and often fraught program to preserve world biodiversity against the encroachments of modern agricultural methods, development, conflict, and climate changes.

I had come to Armenia not to visit ruins, but rather to collect wheat, and even this was an unexpected turn from my research on the history of seed collections in the United States. In the first place, I had been bound not for Armenia but for Aleppo, Syria, then the headquarters of the International Center for Agricultural Research in the Dry Areas (ICARDA). Touted by public relations teams for its location in the Fertile Crescent, the reputed birthplace of agriculture some 10,000 years ago, ICARDA was one of fifteen international agricultural research organizations managed by the Consultative Group on International Agricultural Research (CGIAR), an international public organization funded by the UN Food and Agriculture Organization (FAO), the United Nations Development Program (UNDP), and the International Fund for Agricultural Development (IFAD), among others.[2] These institutes included the Center for Maize and Wheat Improvement (CIMMYT) in Mexico and the International Rice Research Institute (IRRI) in the Philippines, noted for their contributions to international agricultural development. I went to ICARDA because I wanted to study firsthand the international agricultural research organizations so important to the history of world food systems in the second half of the twentieth century, and to explore ways to use seed specimens as sources for writing the history of modern agriculture. But as I studied the herbarium, the gene bank, and the databases that summarized their contents, I became entranced less by the institutions of research than by the objects they organized: the seeds themselves.

So after two weeks in the sweltering heat of Aleppo in July 2010, I boarded a flight to Yerevan with Josephine Piggin, a botanist from ICARDA's Genetic Resource Unit. Two weeks later, in the company of a gene bank manager from New Zealand (Zane Webber), a botanist from the Institute of Botany in Yerevan (Bella Bagratouni), the director of the genetic resource unit at ICARDA (Ahmed Amri), and the Armenian-Russian ICARDA project manager (Natalya Rukhyan), I found myself in a wheat field near a prehistoric archaeological site on a mountain plateau near Sisian.

The area around Zorats Karer is cultivated for local use, so we canvassed the site for old varieties of wheat and its wild relatives: here, chiefly *Aegilops cylindrica,* a modular weed with a cylindrical structure that shatters easily into individual segments. These weeds, which may contain valuable genetic material for breeding, cluster at the edges of a field, as well as on other marginal or disturbed ground such as embankments and roadsides. As the team and I wandered through the ruins and the surrounding fields, we filled paper bags with spikes of some two hundred *Aegilops* plants to be divided among collaborating research institutions.

As I collected in the wheat field, the director of the Genetic Resource Unit at ICARDA, Ahmed Amri, called me over to the crop margin. Among the population of *Aegilops* on the edge of the crop was one plant that didn't look like the others. Less modular than serpentine and twisted, this plant was thicker than *Aegilops* but thinner than wheat. Rather than breaking, it bent. There were no identifiable florets or spikelets as one would find on a wheat plant, even though the plant appeared to be at full maturity.

This plant was neither wheat nor weed. Rather, Dr. Amri explained, it was a naturally occurring cross of the common bread wheat (*Triticum aestivum*) in the field and its wild relative (*Aegilops cylindrica)* on the crop's margin. While most spontaneous crosses, like this one, were sterile, occasionally a cross might produce new seed. That is, the biological process instrumental in maintaining wheat's genetic diversity since its domestication some 10,000 years ago recurs at every moment on the periphery of a wheat field, quite independent of the researchers aiming to systematize it. Although our histories depict agriculture as a practice of human creativity and progress, plants have their own creativity and temporality.

The history of agriculture is the process by which people have attempted to impose order on the generation of plants, attempting to manipulate the rules and habits of other organisms by selecting and modifying them for human exploitation. Naturally self-replicating organisms, seeds nevertheless exist by virtue of human intervention, stewardship, and improvement. Modern agriculture consists of efforts to select and improve plants according to novel rules and systems of organization, including controlled cross-pollination, hybridization, mutation, marker-assisted selection, double haploidy, and genetic modification. The utility and value of seeds hinges on the conditions of their preservation, circulation, and reproduction: laboratory and field, public and private, commercial and communitarian. Seeds are not stable objects, but contested artifacts, classified according to variable logics of science, heritage, and property.

Historians have struggled to situate this long-range development of agriculture within political and natural histories. Plant domestication and migration

Figure 0.2. *A*, Wild cross of *Aegilops cylindrica* and *Triticum aestivum* (common bread wheat) on the border of Zorats Karer. *B*, *Aegilops cylindrica*, a wild relative of wheat. Photos by Courtney Fullilove.

occupy an uncertain position in the oscillation between biological determinism and heroic (or antiheroic) human agency that has characterized global environmental histories. Long-range biohistories offer expansive chronologies of species history and the movement of biological organisms across the globe, challenging conventional chronologies of empire, nation, and capital.[3] Nevertheless, common metaphors of invasion, empire, or colonization employed to describe large-scale biological processes suggest the extent to which concepts of natural science remain mediated by human histories.

Historians of energy and anthropogenic climate change have tried to restore human agency to global environmental history by examining the consequences of decisions about resource use.[4] These histories often reinstate the chronologies of industry and capital species histories have disrupted: the resort to fossil fuels (the industrialization of the late eighteenth and nineteenth centuries) followed the large-scale extraction of energy from the global countryside for the production of agricultural commodities (the global capitalism of the sixteenth to twenty-first centuries).[5] In this seesaw between determinism and agency, cultivated plants act as both subjects and objects of global environmental change.

While crops have moved with human hosts since their genesis, movement and interdependence escalated dramatically after 1500, as European nation-states invaded, appropriated, and integrated new geographies into maritime trade and plantation agriculture. American agricultural expansion in the eighteenth and nineteenth centuries was one manifestation of a *longue durée* of plant transfers resulting from human migration, escalated by European maritime activity from the sixteenth century. The purposeful and incidental transfer of plants from Eurasia to America from the sixteenth to eighteenth centuries supported European settler colonies in the Americas and dramatically altered ecologies on both sides of the Atlantic.[6] Meanwhile, the importation of tropical biota to metropolitan Europe fueled colonial expansion and provided an international infrastructure of nature collection and preservation consisting of ship holds, vented cases, ledgers, naturalists' notebooks, and herbaria.[7]

The merchant-naturalist's quest for useful plants, while providing the machinery of imperialism and colonization, also elaborated a long-standing association between civilization and cultivation. Early nationalists such as Thomas Jefferson turned this impulse to a political vision of an agrarian republic. Drawing on prevalent theories positing a four-stage progression from primitivism to pastoralism to agriculture to commerce, Jefferson and others aimed to fix the moral virtues of agrarian life by securing an agricultural future linked to western lands and European markets. Down the Mississippi River to New Orleans and across the Atlantic, the nation's agricultural surplus supplied industrial Europe with its staple commodities. These political projects, forged against fears of biotic mediocrity and colonial degeneracy, provided the foundation for nineteenth- and twentieth-century models of development rooted in concepts of social evolution and economic growth.[8]

Over the course of the nineteenth century, public research boosted private enterprise through federally consolidated research and development. By the 1880s, the US Department of Agriculture, land grant colleges, and experiment stations pursued improved seeds, mechanization, and chemical applications on the farm. In the economies of scale they supported, farmers filled grain elevators and railroad cars, yoking East to West and producing an agricultural surplus connecting the United States to international markets. Along with rapid industrialization and private property rights in invention, institutions of public agricultural research were handmaidens of capitalist development.[9]

In their attempts to organize improvement, public and private institutions drew on the knowledge and resources of willing and unwilling settlers.[10] This book questions how institutions of improvement exploited, transformed, and displaced extant forms of environmental knowledge and biological material. New seeds and agricultural methods were not so much new as reconfigured, subject to different modes of organization and classification. Agricultural expansion was a story of ingenuity and growth, and it entailed attenuation of biodiversity, erosion of existing craft practices, and co-option of local knowledge for centralized research and development.

The reconfiguration of local knowledge in capitalized research and development abetted some political futures and foreclosed others. Consolidations of biotechnology in agriculture and medicine erased the skilled labor of farmers and pharmacists from the historical record, or relegated them to the status of petty capitalists. The scope and significance of their work indicate a need for greater attention to agrarian knowledge in the ongoing collection of seeds and plants at local, regional, national, and international levels.

A much simpler story of how the United States became a breadbasket to the world remains the centerpiece of conversations about global economic change and the template upon which models of rural development are based. In these potted histories, innovation itself was the foundation of nineteenth-century capitalism, and agriculture was one of its primary objects. The mid-twentieth-century export of American technologies, including improved seeds, transformed world agriculture, diets, and livelihoods. While new scholarship has illuminated how politicos and agronomists provided the ideological and technical apparatus to fashion the developing world after an American image, we understand less about the history of US agricultural development they took for granted.[11]

As producers consolidated regional monocultures of wheat, corn, and cotton, competing visions and economies waned. Modes of resource control originated in plantation agriculture persisted in capitalist forms of labor. Whether interpreted as an artifact of innovation or exploitation, the infrastructure of the global economy consisted not merely of steam engines and fossil fuels but also of fields, fertilizers, seeds, and livestock.[12]

Correlative concepts of natural economy and biological diversity supported agronomic projects fitted to large-scale production. In the twentieth century, new breeding methods consolidated monocultural production and hastened the erosion of agrobiodiversity. Since the 1970s, international agricultural research organizations including ICARDA, CIMMYT, and IRRI have prioritized preservation initiatives to offset these losses, while breeders survey genetically diverse material to produce high-yielding staple crops tolerant of heat, drought, pests, and salinity.[13]

To many critics, however, the large-scale production of cheap food no longer seems secure or desirable, especially to the extent that it supports input-intensive monocultures. Climate variability threatens the viability of water and soil, while economic crisis jeopardizes big agriculture's reliance on cheap labor. In this crisis, seeds are essential inputs as well as variable ones. Food sovereignty movements prioritizing smallholder production and eco-management strategies fitted to family farms may offer a range of alternatives to monoculture and corporate plant breeding interests, but their viability depends on the outcome of broader debates over political economy and resource allocation.[14]

Plant breeding and crop transfers remain embedded in rural development projects. To understand how practices and imperatives of development have changed over time, we need a history of agriculture that shifts focus from institutions of research to the broad field of agrarian knowledge on which they drew. This history should fasten changes in material environments to the systems of knowledge deployed to describe and transform them. And it should identify the contingency and variety of environmental decision making at local, regional, national, and international scales. The more precisely we can identify and understand how practices for the exploitation and manipulation of plants have changed over time, the more effective our political interventions and reforms will be.

Our concepts of preservation should change accordingly. It is intuitive that concepts of nature's economy and useful gardens could sanction the collection and preservation of nature for settler colony, national expansion, and empire. Perhaps less obvious are the ways visions of wilderness or conservation could undergird the same political structures.[15] In each resource-centered concept, however, crisis was visible only to the extent that it affected human social order: through instabilities in commodity prices, human environments, or human health. In the later twentieth century, national governments have invoked vocabularies of security and scarcity to administer food systems.[16] These augment an older imperial lexicon of resources, stock, and treasury and its ethnocentric adjuncts of natives, exotics, and invasions. So captive are we to this imaginary that it's hard to conceive of a style of preservation that eludes specters of threat and endangerment. But as I suggest in the final part of this book, people of the past can help us imagine ways of being and knowing that elevate humility and uncertainty as guiding principles of political and environmental knowledge.

The book is divided into three sections exploring the collection, circulation, and preservation of seeds and plants bound for the United States. The book's action occurs in the neglected corners of two frontiers: the basement of a federal monument in Washington (part 1) and the margins of a wheat field in the Kansas prairie (parts 2 and 3). The storage rooms of the Patent Office and the Post Office were places of accumulation, harboring the detritus of the political economy of innovation as it took shape in the mid-nineteenth-century United States. Seeds were the spoils of collecting enterprises rooted in military, commercial, and scientific expeditions, and as specimens of global nature they were lodged in a crowded museum in the Patent Office Building (chapter 1). The US Patent Office, the institution charged with distributing private property rights in invention, supervised both the museum and the first systematic federal efforts to collect and distribute seeds for the improvement of American agriculture (chapter 2). Its efforts overlapped and on occasion interfered with those of individual improvers, who shared its ambitious and sometimes contentious attempts to envision alternative objects and systems of production (chapter 3).

Part 2 turns from basement storage in Washington to wild grass prairie in Kansas, the ground of agricultural expansion, where local knowledge and resources were impressed into the service of a burgeoning economy. This book takes up the history of German Mennonites emigrating from regions of Crimea and eastern Ukraine, which were branded then, and again in the past two years, as "New Russia." These colonists became dedicated cultivators of hard red winter wheat. National hymns celebrating amber waves of grain as a natural feature of the landscape masked the extent to which refugees and displaced people created the national bounty she celebrated (chapter 4). Moreover, while the Mennonite introduction of "Turkey wheat" to the United States became part and parcel of a mythology of immigrant heritage, this story of hard work and ingenuity concealed a more complex history of wealth, privilege, and property in the making of modern agriculture (chapter 5).

Changing patterns of land use altered not simply the physical environment but also forms of environmental knowledge, including medical practice based on plant drugs. The varied ethics of preservation these alterations inspired are the subject of the book's part 3 ("Indigenous Plants and the Preservation of Biocultural Diversity"). Chapter 6 tracks one manufacturing pharmacist's efforts to acquire a supply of purple coneflower (*Echinacea*) from the diminishing prairie. John Uri Lloyd's efforts, and the decline of botanic medicine's professional status in the later nineteenth century, inspired him to increasingly speculative and baroque reflections on the limits of knowledge systems, reflected in his clandestine service on behalf of the Internal Revenue Service and his science fiction novel describing a journey to the earth's core through the mouth of a cave in Kentucky (chapter 7). If the late nineteenth century could be characterized by a search for order, it was only because of the ultimate irrationality of scientific and legal bureaucracies and the instability of the materials in which they trafficked.[17]

Taking inspiration from the pharmacist's studied labor and muted epistemological crisis, the book concludes by investigating how novel concepts of temporality, continuity, and change applied to seeds have structured modern technological choices, social relations, and modes of production. Inasmuch as their survival requires active stewardship and preservation, seeds embody deep temporal knowledge. Yet their casing renders them opaque, such that they often function as proxies or fetishes, concealing the labor and knowledge they contain. The variable classification of seeds as objects of common use, commerce, and research in the early twentieth century elevated certain actors and institutions and marked the seed's transformation from artifact to commodity.

Interleaved between the book's three parts are field notes from my travels with plant genetic resource specialists in the North and South Caucasus and the Pamir Mountains, which make a case for global histories of grasslands, biocultural diversity, and agrarian knowledge. These interludes are less a plea for the contemporary relevance of the surrounding stories than for a reframing of a national history that has proved both too insular and too arrogant in its stance toward the rest of the world. The Kansas prairie bears striking similarities not simply to the Russian steppe from which many nineteenth-century farmers immigrated, but also to the fields at the foot of Mount Ararat traversed by a Yazidi pastoralist with a nascent interest in dyestuff plants. While political interests continue to structure the exploitation of natural resources, national histories may not be the best containers for understanding environmental decision making. Environmental continuities, and the shared history of environmental knowledge they imply, suggest that if our policies and practices of collecting and improving global plants are blinkered, it is in part a reflection of the histories we have told about them.

So even as I became a collector, I remained a historian.

The title of this book is drawn from the book of Ecclesiastes, a gnomic and tense debate over problems of meaning and accountability, which veers dangerously close to nihilism but settles into a recursive habit of conscience and engagement. The Hebrew of the passages in question is obscure, producing numerous interpretations as to the primacy of political corruption, justice, and divine judgment in human life. "If you see in a province the oppression of the poor and the violation of justice and righteousness, do not be amazed at the matter," the narrator avows, "for the high official is watched by a higher, and there are yet higher ones over them." Highest of all is the king, and "the king himself is served by the field: the profit of the earth is for all."[18] While diagramming the structural nature of exploitation in agricultural economies, these passages also insinuate the universality of sin in a system where all eat by virtue of the labor of others in extracting the produce of nature. Our contemporary preoccupations with scales of production, technological manipulation of organisms, and ownership of natural and intellectual resources reflect the persistence of these moral and political considerations in the organization and administration of world food systems.

Often the succeeding analysis undercuts appealing romances of endangerment, localism, and tradition deployed against technocratic and corporatized practices of modern agriculture, but its intent is to provide more politically useful histories of improvement. Research in the fields of environmental studies, bioethics, and the history of science provide tools to situate the production of knowledge as a collaborative project with willing and unwilling participants. Achievements, profits, and successes necessarily entailed discards, foreclosures, and failures. Not everyone benefited from the organization of agriculture that prevailed. Studying the incremental and checkered legacies of improvement can provide a basis for more aware political and legal interventions and create spaces for the stewardship of knowledge and labor obscured or imperiled by imperatives of production.

Field Notes

"GREEN REVOLUTIONS"

HUNTING TURKEY WHEAT

Two years after my trips to the Tigris and Euphrates and the Aras, I washed up on the banks of a different river: the Neva, in the city of Saint Petersburg, Russia. My task was to research the Turkey wheat, the hard red winter wheat conveyed to the nineteenth-century American Midwest, at its point of origin, drawing on the resources of the institution that sponsored so many early collecting expeditions to the regions of cultivation. My destination was the Nikolai Vavilov Research Institute for Plant Industry (VIR), named for a martyr of early biodiversity preservation. Imprisoned in 1940 for sparring with Trofim Lysenko and running afoul of Stalinist orthodoxies, Vavilov ultimately starved in prison. Lysenko explicitly rejected Vavilov's embrace of Mendelian genetics, favoring Lamarckian ideas of environmentally acquired inheritance. With Stalin's blessing, Lysenko turned his theoretical commitments to a campaign against his opponents. By the late 1930s, Vavilov's theories and habit of consorting with foreigners in the collection of global plant genetic resources had rendered him susceptible to charges of treason.[1]

Yet Vavilov's collections, and the novel theories of genetic diversity and centers of origin for cultivated plants they supported, outlasted Lysenkoism. For much of the three decades prior to his death, Vavilov conducted a series of broad-ranging expeditions, amassing some 250,000 accessions of seeds in Leningrad. According to institutional lore, VIR staff starved in the building during the siege of Leningrad rather than consume the seeds they guarded.

If Vavilov's survey of global seeds provided a model for biodiversity preservation, it also supported a Soviet program of agricultural modernization oriented toward plant breeding for large-scale production. Vavilov served as director of the Lenin All-Union Academy of Agricultural Sciences from 1924 to 1935, and his collections of global plants in those years supported a range of research to produce improved varieties of staple crops.[2] The collection he assembled became the standard upon which international agricultural research organizations would model their gene banks in the later twentieth century, in part because of their orientation toward economies of scale.

Figure FN1.1. N. I. Vavilov Research Institute of Plant Industry, Saint Petersburg, Russia. Photo by Courtney Fullilove.

It was in this storied archive of biodiversity that I pursued the origin of Turkey wheat, with some difficulty. Without a record of VIR's collections, I was left to fend for myself in the database of seed samples. My translator had left the room, and so the curators of the Wheat Department and I sweated nervously in each other's presence, speaking in monosyllables and at a glacial pace. I had come to research Turkey wheat, but it was unclear to them what I meant. I wrote it down on a slip of paper, and the curator furrowed his brow and pulled out his laptop. He tapped on his keyboard for a few moments and turned it to face me. The result was the output from a query of the VIR collections database, showing, indeed, three samples of "Turkey wheat"—all from the US Department of Agriculture. "What is this?" he asked me.

It took a while longer to sort out, as I took the long route toward clarity, sketching a backward path on my slip of paper: "Norin 10," I wrote first, referring to the first semidwarf variety used by the American agronomist and breeder Norman Borlaug in his pioneering production of semidwarf wheat.

Everyone knew Norin 10. The US agronomist Cecil Salmon brought a sample to the United States in 1945, when he was a member of the Agricultural Research Service traveling with US forces in occupied Japan. He was visiting the Marioka Agricultural Research station on Honshu and collected wheat samples from Japanese scientists there. He sent the samples to Orville Vogel, a USDA agronomist at Washington State University in Pullman, who crossed Norin 10 and Brevor, a popular variety in Washington. That cross, Gaines, dominated agriculture in the Pacific Northwest until the late 1960s.

Meanwhile, Norman Borlaug, who had been working in the Mexican Agricultural Program with funding from the Rockefeller Foundation, acquired Norin 10 and its derivatives from Vogel and the USDA. He tried to offset Norin 10's disease susceptibility with American cultivars Daruma, Fultz, and Brevor 14. Eventually he achieved the greatest success with Mexican varieties Lerma-rojo 10 and Sonora 64, which produced the high-yielding hybrids exported to India in the 1960s.[3] These semidwarf hybrids, cultivated with appropriate inputs of water, fertilizer, and pesticides, produce enormous yields, facilitating the technology transfer often referred to as the Green Revolution.

Along with input-intensive agricultural practices requiring extensive chemicals and irrigation, the production of semidwarf hybrids of wheat and rice has been credited with dramatically increasing world food production. This story of origins tends to obscure the parallel pathways of dwarfing genes through Italian, eastern European, and Russian wheat breeding from the 1910s through the interwar period, as well as the primary influence of American political commitments to winning the Cold War by modernizing the global countryside. Coined by USAID staffer William Gaud, the term "Green Revolution" referred implicitly to the "Red Revolution," to be averted by improving the lots of hungry peasants.[4] High-yielding seeds were the symbols of bounty, but it took far more to wage a revolution.

"Norin 10." The wheat curator nodded. Then I drew an arrow and wrote "Brevor 6 → Kanred 6 → Turkey," tracing the variety's parents and their Great Plains ancestors. Turkey was an ancestor of Norin 10, and allegedly it derived from Crimean stock.[5]

We communicated in an international standard of breed names established in twentieth-century research and development, just as in the course of the collecting missions we had learned to communicate in Linnaean binomials, facing each other without a common language and pointing excitedly over an embankment: "*Aegilops tauschii!*"—a wild relative of wheat spotted at long last in a region where we had expected to find it but had not.

But these labels were not fixed to the seeds at germination. Rather, they were selected to designate new varieties, whether according to the names of twentieth-century breeding stations, intrepid plant explorers, or savvy entrepreneurs. Frequently a hybrid of "official" and "folk" names would be reflected even within official nomenclature, as the case of Turkey wheat would demonstrate.

In the field, identification would often be a painful process of translation. A Tajik collector would write a local name for a variety in Pamiri or Tajik, with other notations in Russian. The (Polish) data officer might have rudimentary Russian but lose the Pamiri and Tajik notations entirely. More often than not, a sample would be bagged and logged as simply "red, awned," with no other information. The provenance of the seed was then lost.

"Turkey, Crimea," I said to the wheat curator.

"Krymka!" he exclaimed, answering me with the Russian folk name for the wheat.

We were both hugely relieved at our episode of successful communication.

Vavilov had sponsored some of the first organized collecting missions to the regions from which Turkey wheat was alleged to originate, but few samples remained. The institute retained a handful of accessions of Krymka wheat: one from the second All-Russian exhibition of seeds and machinery in 1912, eight from the Moscow All-Russian agricultural exhibition in 1923, three from VIR's 1923 expedition from Krym University by Vavilov's wife, the botanist E. I. Barulina, and stray samples from the Krym Commissariat of Agriculture, the Odessa Breeding-Genetic Institute, and the Krasnodar breeding station, none of which was dated.

Krymka wheat and Turkey wheat were effectively synonyms, yet the latter was nevertheless distinct: a folk name used by the Mennonites who introduced it to Kansas, which USDA agronomist Mark Carleton picked up when he began researching its properties in the 1890s. When he returned from Russia, he brought with him numerous promising varieties including Crimean hard red winter wheat, separated from the ones first cultivated by the Mennonites by some sixty years. Carleton documented his routes responsibly, but Mennonite settlers brought stocks of wheat from a variety of locations using kinship networks and available commercial channels, and they continued to seek new supplies in the decades after settlement.[6]

So Turkey wheat was Krymka wheat, sort of. But after several generations of cultivation in the United States, the USDA agronomists who collected it knew it as "Turkey"; thus a handful of specimens were at some stage repatriated from the United States to the former USSR. And they had likely changed considerably from the time their Crimean ancestors had traveled to Kansas fifty years before.

The USDA sponsored seed collecting efforts throughout the twentieth century, seeking genetically diverse material to support breeding programs oriented toward economies of scale, and it was in this context that Norman Borlaug conducted his research with the Mexican Agricultural Program in the 1940s. ICARDA and the other international agricultural research organizations of the CGIAR were heirs of Borlaug's research. Aiming to build on the alleged successes of the Green Revolution, the Food and Agriculture Organization of the United Nations supported programs of agricultural modernization and the free exchange of germplasm between countries for the use of breeders.[7]

But there is no consensus that monocultures of scale are the best models for world food production. Climate change and the late twentieth-century globalization of the food supply have provoked renewed concern about the ability or inability of local, regional, and national communities to feed themselves. Human beings survive on a handful of cereal crops. The United States is the fourth largest producer of wheat, behind China, India, and Russia. And it is the biggest exporter of wheat, chiefly to developing countries.[8] Maize, rice, and wheat make up roughly half of the world's caloric intake.

While some credit the Green Revolution with feeding the world, others charge that it ushered in an era of unsustainable practices that strip the soil, exhaust natural supplies of water, and expose workers to dangerous chemicals

Figure FN1.2. Herbarium specimen of red winter wheat from Crimea, *Triticum aestivum* var. *erythrospermum*, collected 1926–27. Photo by Courtney Fullilove.

applied as pesticides and herbicides. Advocates of sustainable development, organic agriculture, local food, and community sovereignty have called into question the wisdom of scaling up production for international commodity markets, offering characterizations of agrarian life at odds with those of input-intensive agriculture.

In the shadow of these developments, I found myself lost in archives at VIR, with three orphaned specimens of Turkey Red Wheat: no one would recognize these as the seeds of the Green Revolution. Modern plant breeding had accelerated processes of natural and artificial selection. As an aspect of this process, researchers, soldiers, and immigrants shuffled material back and forth across oceans. Whether it changed or remained the same, it acquired many new names in its travels.

My lost-in-translation moment betrayed the porous and contingent qualities of biological innovation assembled through proprietary, commercial, and vernacular practices of improvement. I have struggled to recover these practices in an archive while seeing firsthand their erasure in the field and in the archives that sought to preserve their artifacts. One such archive, in the nineteenth-century US Patent Office Building, is the subject of the following chapters.

PART 1 : COLLECTION

THE POLITICAL CULTURE OF SEEDS

(Patent Office Building, Washington, DC; Fujian; Assam; South Carolina, 1842–59)

Figure 1.1. Patent Office Building, 1846, daguerreotype by John Plumbe Jr., with greenhouse constructed to house collections of US Exploring Expedition to the Pacific (1838–42) visible to the rear right. Library of Congress, Prints and Photographs Division, LC-USZC4-3596.

1 : THE MUSEUM OF SEEDS

As Commodore Matthew Perry sailed his gunboats to Edo Bay in 1853, charged with compelling Japan to trade with the United States, the expedition's agriculturalist, James Morrow, weathered the passage to the Indian Ocean in the hold of a store ship, trying to keep his plants alive. The plants and papers of garden seed, provided by Philadelphia horticulturalists and the United States Patent Office, were intended as diplomatic gifts to support Perry's mission. Morrow took his charge seriously. In gale force winds, he tried to prevent the plants being doused with salt water and spray, but rather than the state-of-the-art Ward cases made of glass, Morrow had old ones from China outfitted with oyster shells as vents. In storm after storm, the shells broke, the tarpaulin blew off the case, and the plants were doused in seawater, parching some of the hardiest.[1]

Once on shore, Morrow spent happier days in country rambles, collecting seeds and cuttings for the use of American farmers, gathering whole plants he dried and pressed into herbarium specimens back on the ship at night, and attempting with some difficulty to purchase examples of agricultural implements not used in the United States. In Okinawa, he tried to buy a plow that caught his fancy, only to have half the village claim partial ownership. As the negotiation stretched upward of an hour, Morrow faced his translator in consternation, wondering how local agriculturalists could be so admired and yet too poor to afford a plow.[2]

Morrow assumed the plow was singly owned rather than subject to overlapping entitlements, reflecting a simple and perhaps nationalist conviction in rights of private property. Yet in reality, there was nothing simple about American concepts of property. The seeds, dried plants, and agricultural implements Morrow collected would all be classified and distributed differently upon their return to the United States. Morrow sent his herbarium samples to Harvard's Asa Gray for identification; they survive as specimens of global nature in the collections of the US National Herbarium, New York Botanic Garden, and the Natural History Museum in London.[3] The plow Morrow had worked so hard to acquire in Okinawa went on display in the Patent Office museum with the other diplomatic gifts and Japanese handicrafts collected by the Perry Expedition, ultimately forming the kernel of the ethnological collections in the nascent US National Museum of the Smithsonian Institution.[4]

Meanwhile, the seeds and cuttings went to the Philadelphia horticulturalists who had helped supply the expedition, and to the US Patent Office for gratis distribution to interested farmers. In transit, they remained in storerooms and greenhouses adjacent to the Patent Office Building. Although we can follow the progress of many objects through the Smithsonian or dispersed herbaria, the seeds and cuttings Morrow procured are the most difficult to track. Museums with aspirations to immortality prioritized permanent display, which in turn required inorganic objects, or feats of preservation arresting decay. In contrast, seeds require regeneration to remain viable. And once distributed to American farmers, they became the property of their cultivators, not the federal government.

Did seeds have value as commodities or scientific specimens, and what rules of exchange governed their transfer? The ultimate disaggregation of specimens into separate institutions—museums, botanic gardens, seed companies, and private farms—has concealed the many types of collection, appropriation, and exchange that occurred within diplomatic contexts, as well as the varied rights of ownership applied to the objects acquired. In museums, things became material signifiers of cultural difference, interpreted according to theories of social evolution.[5] Herbaria in turn represented an insistence that global nature was universal and to be shared, although such collections often obscured the hierarchies of human labor and systems of knowledge that sustained them.[6] Morrow's project was not to contribute to a universal store of knowledge, however, but rather to profit American agriculture.

For a time, these varied collections jostled in the halls of the Patent Office Building. As the only federal agency charged with managing problems of knowledge in the early republic,[7] the Patent Office became a theater for conflicts over value and custody. Its mandate "to promote the progress of science and the useful arts" derived from the constitutional clause establishing patents and copyrights. But the exact scope of the mandate remained undefined, alternatively oriented toward property rights in invention, exploration, research, publication, and expedition. These conflicts over the proper political economy of knowledge in the early United States expressed themselves in different claims to the specimens that flooded the halls of the Patent Office Building. Rather than staging a system of classification shared by curators and spectators, the Patent Office museum harbored an array of contradictory approaches to the

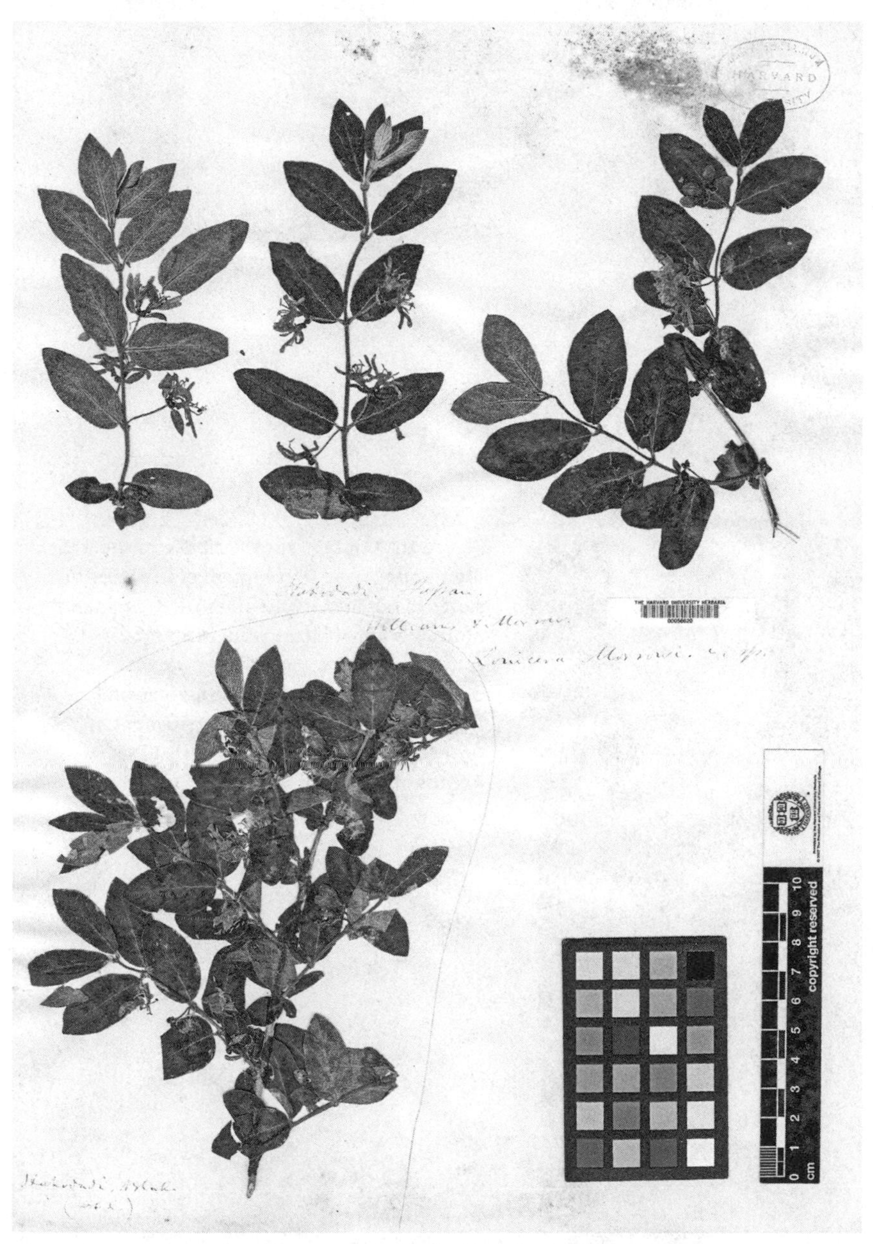

Figure 1.2. Holotype of *Lonicera morrowi* in the Gray Herbarium, Harvard University. Asa Gray christened this honeysuckle bush in the name of its collector, James Morrow. Gray's lengthy delay in identifying the Japan Expedition's botanic specimens for publication aggravated Commodore Perry. Duplicate specimens are lodged in the herbaria of the New York Botanical Garden and the National Herbarium, Washington, DC. Gray Herbarium.

Figure 1.3. George Jones, chaplain of the USS *Mississippi*, used Department of the Interior funds to purchase these hoes (*A*, *B*) on James Morrow's behalf at a bazaar in Shimoda following the Treaty of Kanagawa. Morrow describes the bazaar and the agricultural implements in his journal. Department of Anthropology, Smithsonian Institution. Photos by James Di Loreto.

Figure 1.4. Plowshare and blade made of cast iron acquired during the Perry Expedition and placed on display in the Patent Office museum, now in the collections of the National Museum of Natural History, Smithsonian Institution, Washington, DC. Department of Anthropology, Smithsonian Institution. Photo by James Di Loreto and Donald E. Hurlbert.

material it housed. Naturalists' collections, agricultural fairs, and mechanics' institutes all served as models for the exhibition of artifacts in the museum, and competing visions meant competing systems of organization and display.

Meanwhile, tourists and journalists flocked to the halls of "the great cabinet of curiosities" in the Patent Office Building, which opened to the public in 1849 and quickly became the capital's most popular tourist attraction.[8] Alternately styled as a cabinet of curiosities, a national gallery, a temple of invention, and a rational place of amusement, the Patent Office galleries amassed specimens of natural science through the navy and consular service, American Indian artifacts symbolizing territorial mastery over the continental West, and donations from American manufacturers and agriculturalists. Collection relied on practices of territorial and economic expansion: naval expeditions, the violent removal of American Indians and annexation of Mexican territories, and the development of circuits of commerce for agricultural products and manufactured goods.[9]

Scientists and statesmen hashed out the meaning of science and useful arts through turf wars over appropriations, rooms, and collections housed in the Patent Office museum, which were ultimately transferred to the Smithsonian Institution when it was inaugurated in 1857. In the Patent Office, and variously bound for greenhouses and storerooms, plants were subject to the debates over meaning and value. The museum remained a site where problems of knowledge went unresolved, revealing a state that was less inchoate than disordered, administered by politicians, bureaucrats, and foot soldiers with competing visions of proper government and public knowledge.

Ultimately, disputes over the value of specimens were less about prices, as Commodore Perry would have had it, than the character of global connections being forged. For while the decade before the American Civil War was a moment of expanded commerce and global aspirations, it also entailed varied imaginations of the global: as a patchwork of militarized nation-states, a hierarchy of civilizations, a grid of marketplaces, and a zone of common nature. Each model required different rules of conduct, which often contradicted one another in theory and practice. Is trade free that requires a threat of force? Do people need permission to buy, barter, or take? What if the goods in question are products of nature rather than human labor? How should ownership and value be determined across societies with different organizations of property and worth? The persistent collection and exchange of global seeds and plants provoked these questions.

In spite of their ultimate obscurity, seeds and plants were the most common collections in early national museums constituted of global military and scientific expeditions. In 1843, Secretary of War John M. Porter had advocated putting a pine box on every outgoing vessel for collections of nature specimens, charging every officer of the navy to devote his free time to the increase and diffusion of knowledge.[10] The collecting enterprise had begun on a grand scale with the Wilkes Expedition to the Pacific (1838–42), and it continued with naval expeditions to the Dead Sea and the River Jordan (1847–1849), the Herndon-Gibbon Expedition to the Amazon (1851–52), the Naval Astronomical Expedition to the Southern Hemisphere (1846–52), the Page Expedition to Rio Paraguay and Rio de la Plata (1853–56), and Commodore Matthew C. Perry's 1852 and 1854 expeditions to Japan.[11] Alongside the spoils of naval expeditions were collections from the Creek and Seminole Wars, the Mexican War, and Charles Fremont's collections with the Corps of Topographical Engineers in the American West. Seeds and cuttings choked storage rooms and public galleries of the US Patent Office and sometimes disappeared in the purses of powerful visitors with horticultural fancies.

Proponents of American expansion targeted naval reach and continental settlement. The Pacific figured as a special object of exploration and commerce, with products of nature moved overseas in the holds of merchant brigs and naval warships before being wheeled into the basement corridors of the

Patent Office Building. This exercise in collecting was consistent with, and indeed modeled on, European imperial exploration of the preceding three centuries. The continental West presented a field for collection and documentation, with evidence of native technics used to support claims to the superiority of Euro-American institutions of science and property.

The Patent Office museum followed the pattern of early national museums in Europe, many of which derived from private collections and cabinets of curiosity enlarged through colonial expansion, imperial acquisition, scientific exchange, and public and private patronage. Like the museums to which Americans looked for inspiration, the one in the Patent Office Building expressed national power, addressing its own citizens and the elite members of other nations. It nested claims to education and enlightenment within an ideology of national competitiveness and commercial might.[12] In the machinery of expeditions and surveys, scientific societies, agricultural improvement, and patents for manufactures, the Enlightenment imperative for the increase and diffusion of knowledge and the political imperative of expansion met.

The Patent Office museum differed from its European precedents in part because of its belated and derivative formation.[13] Like many museums, the one in the Patent Office Building was a site of conflict between diverse constituents, interests, and claims to authority, each of which expressed different ideas of the functions the galleries should serve. But rather than legitimizing an established state, the museum in the Patent Office Building expressed all the indeterminacy and conflict of state development itself.[14]

No one felt his disunity more keenly than the museum's caretaker, John Varden, who went with his personal cabinet as custodian after it was acquired by the leadership of the National Institute for the Promotion of Science, then attempting to build a cabinet in a bid to acquire James Smithson's bequest to the United States, "to found at Washington, under the name of the Smithsonian Institution, an Establishment for the increase and diffusion of knowledge."[15] Varden was a set designer for traveling theaters with a penchant for collecting, gathering American Indian relics and assorted nature specimens from New Orleans and Mississippi and throughout the mid-Atlantic region. Through collecting and exchange, he had amassed sufficient material by the late 1820s to open his collection to the public.[16] A year later, it was acquired by the National Institute.

Varden had not one but three bosses over the course of a single decade. After the National Institute acquired his cabinet, Varden's collections went to the upper floor of the Patent Office Building, which was then the largest building in Washington and the plausible materialization of the first Superintendent of Patents William Thornton's dream of a "National Museum of the Arts" modeled on the Conservatoire National des Arts et Métiers in Paris.[17] Thornton's successor, Henry Ellsworth, had other ideas, soliciting donations from manufactures and agriculturalists for display in the upper halls. Ultimately, Ellsworth was compelled to share the Patent Office Building with the National Institute, which made a case for appropriations to manage the collections of the scientific corps of the US Exploring Expedition. The expedition was then winding its way through the Pacific under the direction of Charles Wilkes, and Wilkes, too, made claims on Varden's time.[18]

Varden used different titles for the gallery depending on whom he was addressing. To Commander Charles Wilkes of the US Navy, who promoted the botanical, ornithological, and ethnological spoils of his Pacific expedition, it was the Hall of the US Exploring Expedition. To the members of the National Institute, which contributed its own scientific cabinet assembled from donations, it was the National Gallery. To Commissioner of Patents Henry Ellsworth, who solicited donations representing American ingenuity, it was the National Gallery of Manufactures and Agriculture. To everyone else, it was the national cabinet of curiosities, obviating the more specific visions of the museum's progenitors. Meanwhile, ever more plants, seeds, and cuttings made their way to US shores, setting the stage for a crisis over the meaning of the specimens clogging the halls of the Patent Office Building. Were seeds scientific specimens, objects of common use, or commodities?

The exchange of seeds as diplomatic gifts showcased the simultaneous and contrary values of global nature, private property, and tool of commerce applied to them—contradictions that persisted in subsequent institutions of research and development. In the absence of the market norms for international exchange Perry tried to impose, Morrow struggled to fix the value of seed specimens variably determined by customs of gift, barter, and market exchange. What determined value? Morrow negotiated these questions on the ground, with limited success.

Officially, Morrow and Perry were bound for Japan on a diplomatic errand. When Perry first docked in Edo Bay in July 1853, he intended to deliver a letter from President Millard Fillmore to the emperor. While Japanese officials attempted to redirect him to the port of Nagasaki, where the Dutch had limited and exclusive rights of trade, Perry would not be moved. After accomplishing the delivery of the letter on his own terms, he retreated to Macao to provision his ships and await the shipment of gifts intended for the emperor, which were mentioned explicitly at the conclusion of Fillmore's letter.[19]

The store ship *Lexington* reached Hong Kong the day after Christmas bearing the official gifts for the Japanese, including a quarter-scale fully operational railroad, a telegraph set with three miles of wire, a set of standard weights and measures, navigational charts provided by the US Coast Guard, and a complete Double Elephant folio of John James Audubon's *Birds of America*. In the interim, the sloop *Vandalia* arrived as well, bearing numerous agricultural implements, seeds, and plants also intended for distribution as gifts, and with them James Morrow, the South Carolina physician and agriculturalist charged with their care. Morrow had accompanied the *Vandalia* from Philadelphia to Rio de Janeiro and the Cape of Good Hope onward through Java and Singapore to Hong Kong and Macao, surveying agricultural practice and collecting seeds and plants at every stop.

Morrow's mandate was to distribute and collect plants on behalf of the US government. Secretary of State Edward Everett's marching orders put Morrow under Perry's command and instructed him to "take charge of the Seeds and agricultural Implements procured for the Expedition," including stocks from the US Patent Agricultural Department and the private firms of David Landreth Seed Company and Robert Buist and Company, both in Philadelphia.[20] The latter provided numerous small papers of garden seeds for Morrow to distribute as gifts in his travels. In addition to superintending any agricultural operations ordered by Perry, Everett instructed Morrow to use the seeds to introduce new crops into the places he visited, and to "carefully note and collect all indigenous vegetable products, within your sphere of operations, with a view to their introduction into the United States." He also instructed Morrow to keep a journal of his activities for the records of the department and preserve "seeds and dried specimens of as many plants as possible."[21]

Morrow collected by all the usual means. At Rio he purchased from a market opposite the square from the landing, noting that mules rather than wagons carried the produce to market over the rough mountain roads. When Morrow asked the market vendor about unfamiliar vegetables, she invariably responded with only the price, assuming everyone had knowledge of the articles themselves. He looked for seed stores but found none; almost everywhere, seeds were not market commodities but exchanged locally between farmers. The consul in Brazil presented him with specimens of manioc flour and other plants. In many places he visited, he marveled that most cultivation was performed by hand. In Brazil, he watched workers bearing cured blocks of India rubber sap down from the mountain in blocks.

When Morrow assessed plants for their utility, he was generally agnostic as to the proper political economy of agriculture in the United States. But he devoted special attention to all staple crops, including those used to support slave plantations. In Java, he described rice in three varieties at the water's edge: large and yellow grained, small and purple, and bearded. The farmer provided samples of each but demurred when asked from where the rice came and how to go there. While Perry was negotiating the delivery of Fillmore's letter in July 1853, Morrow was touring sugar mills and nutmeg plantations with the US consul in Singapore, noting the recent introduction of nutmeg and cassava as well as the predominance of Chinese labor and agricultural implements.

Throughout his collection, Morrow drew on knowledge acquired growing up on his stepfather's plantation in South Carolina. In Rio, Morrow purchased black beans from the grocery because of their use with jerked beef as slave provisions in the American South. In Singapore, he observed bales of American cotton at the wharf and attributed recent failures in its culture to the ignorance of cultivators, poorly suited soil, and a rainy season that rotted the bolls before the plant reached maturity.[22]

Morrow's experience as an agriculturalist gave him a different perspective on the places he visited, perhaps less politically naive than Perry's determination to find ports of refuge for American steamers. Before setting sail, Perry had identified Lew Chew (or the Ryukyu Islands, now Okinawa), approximately eight days' sailing to Edo, as the best base for expeditions to Japan. He boasted to Secretary of the Navy John P. Kennedy of his success in earning "the good will and confidence of the conquered people" when obliged to "subjugate" many towns and communities during previous commands on the coast of Africa and the Gulf of Mexico.[23]

Perry imagined the islands as a base of supply for American vessels and aimed to promote the introduction of American crops for this purpose. For Perry, seeds were tools of international commercial development. He recommended "encouraging the natives in the cultivation of fruits, vegetables, &c.," by means of the garden seeds provided. He also suggested providing agricultural implements as gifts, including plow and harrow, spades and hoes, threshing and winnowing machines, and especially gins for cotton and rice. Ultimately, Perry directed his efforts east of Lew Chew to the more sparsely inhabited Bonin Islands, where he followed through on his plans. Perry regarded the islands as well positioned as a stopping place for mail steamers and other vessels likely to traverse the Pacific in coming years. At Peel Island, Perry's officers oversaw the signing of a treaty among settlers and residents establishing a colony characterized by basic rights of property and relations with docked ships.[24]

Perry's tabula rasa imagining of Okinawa's agricultural future might have amused Morrow had he known of it; when Morrow visited the islands en route to meet Perry in Macao, he found that "every foot of ground" appeared to be "carefully cultivated." Even as the ship approached Naha, he admired the "various shades of green presented by the different winter crops" as well as the "uniformly terraced hills" and groves of trees forming a "most beautiful cultivated rural landscape." He admired the carefully trimmed hedges characteristic of all villages and the broad, smooth roads with ditches flanked by rows of pines.[25]

Morrow admired the careful agronomy of Okinawan cultivators. Sweet potatoes and kidney beans were cultivated in beds simultaneously. Turnips, radishes, and parsnips grew irregularly where potato vines were sparse or missing. Raised beds allowed cultivation on rice lands. On mountainsides, banks were planted with ferns and carved with small drains to prevent the soil from washing away in heavy rain. He admired the water management in the rice fields, constructed for ideal reservoir and drainage and the extensive sugarcane cultivation in the middle of the island.[26]

Here as elsewhere, Morrow paid special attention to staple and cash crops. Wheat, millet, and barley were cultivated throughout the territory, the subsoil scrupulously prepared with hoe and compost. Morrow noted the cultivation of tobacco and cotton, and smooth-headed and bearded types of wheat somewhat afflicted by rust and Hessian fly. He admired the peculiar reverse pyramid structure of their granaries, designed to keep out moisture and rats, and supposed the granaries managed by town or government according to population density rather than cultivation area.[27]

Morrow's study of agriculture included systems of labor. Cultivation was by hand with limited domestic animals; implements were principally a wooden harrow and an iron chisel. Morrow thought the prevalence of labor made the introduction of agricultural implements impractical, remarking that the population appeared "happy and contented" in spite of being "under the worst form of slavery." Morrow, who five years later would be a staunch defender of the Confederacy, nevertheless showed some attention to conditions of exploitation. "The women," he observed in passing, "are slaves to slaves, in being the slaves of the men."[28]

Meanwhile he set about procuring seeds, cuttings, and implements of interest, including the plow he labored to secure for the collections of the Patent Office. "After a long consultation with the ploughman and a great deal of talk among themselves," Morrow recalled, "they fixed on a reasonable price," and "we moved off to the village with our purchase, followed by about twenty men and women." On reaching the village, Morrow unfortunately encountered the owner of the plowshare, the most valuable part of the plow. "The plough and the gearing were owned by four different persons," Morrow marveled, "the nose-ring and line by one; the yoke, traces, and single tree by another; the plough stalk by a third; and the share by a fourth."[29]

In marveling that cultivators with such a reputation for skill were too poor to afford a plow, Morrow sidestepped more fundamental questions about the organization of agricultural production in both Okinawa and the United States. On his stepfather's plantation in South Carolina, chattel slavery supported planter wealth, and in the western territories, homesteaders acquired implements through indebtedness. Regardless, Morrow's task was confined to technical observation on agronomy and the collection of useful plants without regard for the systems of labor they entailed, a tendency that has proved a persistent aspect of twentieth- and twenty-first-century agricultural science.

For Morrow, seeds were both official gifts and tokens of appreciation, but he also struggled to fix their value as commodities during his awkwardly administered tenure under Perry's command. When Morrow ultimately reported to Perry in Macao at the end of August, Perry had no information of his post. Secretary of State Edward Everett had appointed Morrow as expedition agriculturalist after Perry had already sailed. Perry nevertheless determined to put Morrow to good use, reviewing his letters of introduction from Everett and the president of the Philadelphia Horticultural Society and summarily instructing him to request funds from the secretary of the navy for the purchasing of seeds and agricultural implements he had been charged to collect. Perry loaned Morrow money from the expedition's funding. When the squadron returned to Kanagawa from February to April of 1854, Morrow took up his charge in Japan.

Wherever Morrow went, he gave away garden seeds and plants as gifts to those who helped him. He often gave garden seeds to expats and consuls, but also to local farmers who purchased or gave seeds to him. Morrow visited numerous fields and villages during the squadron's months on shore. Toward the end of the squadron's time at Kanagawa, he returned to several villages up the valley he had visited previously. Farmers recognized him and expressed gratitude for the seeds he had brought, which they had planted. Morrow gave them more seeds as a farewell.[30] The missionary interpreter Samuel Wells Williams, who was a naturalist by hobby, often accompanied Morrow, collecting many plants in the wild and making observations of the local flora. When invited for tea and refreshments at local houses, Morrow gave his hosts the papers of seed he had brought with him.

Morrow also gave seeds to the Japanese officials who helped him. In addition to various military officers, the local governors at times accompanied Morrow and Williams on their walks. On one walk with the regional governor and two youths, the boys helped him dig up small pines and a peach tree. Morrow observed that they seemed to understand the purpose for which the trees were intended. Yezaimen, the governor of Uraga, also requested turnip seed from Yokohama for Morrow's collections. In return, Morrow gave him several papers of seed. "Although he resides in the city of Yeddo," Morrow added, "he seemed very grateful for them, as does every body in Japan."[31]

Seeds had special status as gifts, in that they seemed to be the only thing people were allowed to receive without hindrance. Morrow reckoned the exception was strategic: "the importance they attach to everything connected with agriculture is so great, that they do not seem to have included this [seeds] in their strict prohibitions to foreign intercourse. Bread or meat or any other simple thing would have been returned had I given it to the people,—but everybody is eager to receive seed."[32] This exception for seeds extended to official gifts from the squadron to Japanese officials, which were otherwise subject to severe restrictions. Assistant purser William Speiden noted that Yezaimen received "a quantity of choice garden seeds" even before the larger exchange of presents in the evening. Yezaimen and other officials also received large boxes of American seeds conveyed by Morrow.[33]

In fact, seeds were so valued that they were often taken from the recipients. Morrow lamented that local officials often took from common farmers the seeds he had given as gifts, not because they were proscribed but because they were useful. "I am sorry to learn that some of the seed I had distributed to the common people had been taken away by the officers above them," he noted, "not to return to me, as they do everything else, but from their own cupidity, to use it themselves." He further objected that such officers were slow to give him seed in return."[34] Williams, too, hoped a favorite host who had accompanied them on walks would be able to keep the gift Williams gave him in parting. The man admitted he had been "forced to give up some of the seed formerly given him."[35]

In spite of Morrow's complaints about the cupidity of officers relative to common farmers, he did note organized efforts to collect seed on his behalf. In addition to the small bag of thirty varieties of garden seeds allegedly sent by the emperor from the Imperial Gardens of Edo, one high officer Morrow described as an "accomplished scholar" assembled a parcel of fifty varieties in somewhat greater quantities: "no doubt with great pains," Morrow noted. The man explained that there were no stores from which seeds could be purchased, as they were "distributed, by exchange, from one neighbor to another," a practice that was customary in many places Morrow visited, and in the United States. The noncommercial means of exchange nevertheless rendered collection a slow and complicated undertaking.[36]

Seeds and agricultural implements also formed part of the official gift, a display of force intended to impress the Japanese both with American technological advancement and determination to trade. The assembly and presentation of the varied gifts would stretch upwards of a week, during which time Perry continued to negotiate the terms of the treaty with government officials.

On the appointed day of delivery, Morrow woke early to prepare his portion of the gift on the store ships *Southampton* and *Lexington*. In spite of heavy rains, ultimately six ships stocked with gifts crossed rough waters and landed safely, carrying crates of arms, potatoes, lifeboats, books, whiskey, seeds, and all the pieces to assemble the miniature railroad, telegraph, and agricultural machinery, all of which were offered as examples of American manufactures. Morrow wiped down the agricultural implements drenched in the storm. Over the next week, Morrow continued to unpack, arrange, and assemble the agricultural implements, and then to demonstrate their operation to the Japanese.[37]

After the treaty had been signed, the squadron set out to test it; for his part, Morrow set about procuring as many seeds and plants as possible at the new treaty ports. He made purchases both in the official bazaars opened to Americans and from local farmers and gardeners. At the first market in Shimoda, he purchased a bag of black wheat. He had more trouble buying rice, which farmers resisted selling. This was not unique.

In many stops along the route, in Japan and elsewhere, Morrow had trouble buying seed. This was in part because seed was not a saleable commodity; farmers traded seed among themselves. But some farmers also expressed outright reluctance to sell seed grain, a widespread resistance indicating some proprietary tendency among farmers to protect seed they had selected and improved over time.

When Morrow was not able to procure seed grain from farmers, he approached the chief officer at the customhouse for help. This yielded limited results. In spite of requesting a full sack, he was able to procure only very little at a high price. Ultimately, when the fleet departed Japan, the customhouse sent barley and wheat Morrow had requested, but he was dissatisfied with the lot, which he had no ability to select himself.[38]

Meanwhile, Morrow began to use his rambles not simply to acquire seeds and plants by purchase, gift, and collection, but also to investigate the market prices of items beyond the official bazaar. Tariffs and other charges there inflated prices. When Morrow went to stores outside the bazaar, he was often able to buy at local prices. He developed a strategy of observation, taking a seat and watching merchants with other customers.[39] This only increased Morrow's irritation at the inflated prices offered to Americans.

Over the course of the spring and early summer, Morrow and others became frustrated with restrictions, prohibitions, and manipulations of price by customs officials. They found the sale of many items prohibited. In Hakodate, Morrow experienced many restrictions on his movements beyond the official market. When he attempted to buy specimens of cotton in local stores, he was refused and redirected to the bazaar. According to the terms of the treaty, moreover, sellers could not receive money directly, but had to accompany buyers to the inspection and customhouse. There sales were often canceled.[40]

Disputes over value did not end when the plants departed Japanese shores. The frictions and uncertainties Morrow experienced in the field paralleled a crisis over the value of collections in the Patent Office Building, where seeds and cuttings alternately figured as scientific collections, raw materials for agricultural improvement, and objects of fancy for horticulturalists. As the Perry collections and numerous gifts and specimens of industry crowded the Patent Office museum, the oversupply and multiple rationales for collection and exhibition precipitated a crisis over the value of specimens, including the seeds and plants. Even the scientific value of collections was a matter of some dispute, with objects alternately representing sensibilities of curiosity, gentlemanly community, and natural type.

For John Varden, specimens were valuable as curiosities, singular representations of natural diversity. He was closer to the collections than anyone, unboxing, documenting, preserving, and arranging them for storage or display. Changes in his documentary practice nevertheless indicated broad changes in the values applied to specimens. As objects moved from the custody of the National Institute and the commissioner of patents to the Smithsonian, lack of documentation compromised the value of his specimens. While he made several inventories of his own collections, he struggled to identify the origin of many objects he had acquired before 1829. In one instance, he recorded the possession of one "Hair Brush made by the natives of [______]." In others he recorded the provenance of a donation—for example, ostrich eggs and a porpoise's jaw bone from Baltimore—without information about its original acquisition.[41] Varden's attention to geographic origin and provenance was an indication of a shift away from curiosity as a primary focus. Rather than being valuable because of its self-evident singularity, the specimen became a host for information about geography, environment, and culture.

For the officers of the National Institute, specimens materialized a vast corresponding network of scientifically inclined men. According to the National Institute, its collections were meant to serve "every department of science and art, the study, investigation, or history of which" could be "aided by such accessories." But with collection based on copious and unscreened donations, the institution was flooded. In a short time, it "found itself in possession of many rich and rare collections, and of many specimens, which though not rare, are nevertheless highly useful and greatly valued."[42]

This formulation implied a hierarchy of value, the criteria of which remained unspecified. Richness and rarity accrued to some specimens, though these qualities referred simultaneously to multiple regimes of value. In terms of free market economy theory, an object's worth increased in proportion to its scarcity; the market price of a commodity increased if demand exceeded supply. Yet rarity could also imply noneconomic values of singularity associated with nature's curiosities. Viewed in this light, rarity was important not as a mark of scarcity, but because it expressed diversity and abundance.

While the ambiguity between natural and economic rarity was unresolved, both measures of value presented problems for the leadership of the National Institute. The institute denied all attempts to monetize the value of collections, rendering market logic inapplicable. In its plea to Congress for appropriations to care for its collections, the institute declined to represent its collections in dollars and cents. Instead it asserted simply that "it would be impossible to purchase many of the specimens at any price."[43] In fact, the commodification of specimens was not impossible, however unpalatable to the leadership of the National Institute, which regarded them as a treasure of science, subject to rules of exchange based on civility and reciprocity. Moreover, given the glut of materials, most of the National Institute's collections failed to meet any definition of rarity.

According to the institute, these extensive but common collections were nevertheless "useful" and "valued." Although they specified no criteria for either, the leaders of the National Institute prized specimens as aids to the study of the natural world. Yet as advances in printing technology enabled the diffusion of scientific knowledge, arguably a single public exhibition in Washington was no longer essential.

To be useful for study, moreover, specimens had to be accessible through either exhibition or exchange. Eighteenth- and nineteenth-century networks of natural science operated according to hybrid rules of exchange derived from customs of gift and commerce, in which collectors donated specimens with the expectation of receiving credit, information, or other specimens. Sale usually relinquished rights to a specimen, whereas donation could imply continued intellectual privilege and control. Liberality with specimens indicated worthiness as a scientific correspondent. Above all, the donation of specimens conferred social status and scientific authority, albeit according to the strictures of a largely preordained social hierarchy.[44]

Yet the National Institute's collections could not be exchanged, calling their value into question yet again. When the Library Committee was granted custody of the US Exploring Expedition collections in 1842, the National Institute lost its license to exchange duplicates for novel specimens. Almost immediately, the leadership of the institute complained that efforts to exchange had been "paralyzed, for want of this right."[45] Although its protests fell on deaf ears, it continued them periodically for the next five years as the institution gradually unraveled.[46] Meanwhile, for want of space, the vast majority of its collections were not exhibited.

In Congress, the glut of mismanaged collections raised questions of ownership. Whose property were the specimens? Congress considered giving duplicates back to the collectors, primarily the scientific corps of the US Exploring Expedition. This proposal aggravated the leadership of the National Institute, which argued that returning specimens to the scientific corps would disrupt the prevailing system of reciprocity.[47] This reasoning was disingenuous. In fact, the leaders of the institute simply wanted control over the terms of the exchange.

The exchange-based economy of scientific prestige was not the only corpus of value applied to specimens in the Patent Office museum, however. Commissioner Ellsworth, among others, interpreted them in terms of their use value as the raw materials for agricultural improvement. He had only acquiesced to hosting the collections of scientific expeditions in exchange for agricultural appropriations to continue the seed exchange and distribution programs he initiated in 1836. In 1839, he recommended the exhibition of the plants and seeds in the Patent Office, anticipating the great contribution to their collection made by the Wilkes Expedition.[48] Soon after, he received official appropriations to continue his collection for the benefit of American farmers. Seeds and plants intended for the Agricultural Department arrived alongside collections for the National Institute, frequently from the same expeditions.

Ellsworth interpreted the Wilkes Expedition as a boon to American agriculture, but he could not have been prepared for the over 60,000 plant and bird specimens returned by the expedition or the vast ethnographic collections that accompanied them. Soon he complained of the dominance of science collections in the gallery. He had no objection to the idea of the Patent Office as a public exhibition space, nor was he opposed to the display of scientific specimens. Rather, he had a specific idea about what made a specimen valuable. Collecting expeditions were instrumental to the promotion of American agriculture, which he interpreted as the core of the exhibition and part of the Patent Office's larger mission to promote science and the useful arts.

As a scientific community jostled with a horticultural one, quibbles over the custody of collections indicated further disagreement over what imbued a specimen with value. While collectors viewed their specimens as contributions to the botanic record, Ellsworth modeled his efforts on traditions of seed sharing and mutual aid, sanctioning the global appropriation of natural resources in the promotion of American agriculture.[49] Ellsworth's model was not simply territorial. It implicitly criticized practices of natural science collection. They implied one could know about the world by amassing and classifying its products, but this practice conveyed little about human creativity. His shelves of buttons and rubber and giant corn roots, on the other hand, established that human beings could transform nature to useful industry.

The establishment of the Smithsonian in 1846 effectively ended the National Institute's hopes of securing the Smithson bequest. Soon, competing organizations challenged the institute's bid for power.[50] Senator Benjamin Tappan of Ohio, chairman of the Library Committee, supported Charles Wilkes in his challenge to the institute's bid for absolute control of his expedition's collections. Tappan's indictment of the institute's authority may have been personally motivated. His habit of "borrowing" the expedition's specimens for his personal garden had earned the ire of those charged with its care.[51]

The problem was not that specimens were meant for glass cases rather than circulation; the institute had lobbied aggressively for the right to exchange duplicates, especially seeds and cuttings. But Tappan didn't follow norms of exchange for natural science; he simply treated the collections as anyone's (or at least his) for the taking. And after all, the bulk of seeds and cuttings collected in government expeditions went through the Patent Office's Agricultural Department, not the National Institute. Who was to say which plants were there for free distribution and which weren't? And at any rate, weren't seeds something shared freely rather than bought and sold?

For his part, Secretary of the Smithsonian Joseph Henry had never wanted a museum—but he got one anyway. Henry grudgingly accepted the government science collections in the interest of securing appropriations for the institution's research and publication exchange programs. In 1857, the Wilkes collections moved from the Patent Office Building to the Smithsonian Castle, followed shortly thereafter by Smithson's personal effects and the other collections of government science expeditions. Meanwhile, in 1861, the newly formed US Department of Agriculture took custody of the seeds and cuttings stored in the Patent Office and the nearby propagating garden. Removed to other facilities, the seeds became invisible as museum objects, and subject to new rules of property and exchange.

In succeeding decades, new taxonomies for botanic, ethnological, and historical relics reconstructed useful knowledge according to the purportedly objective, apolitical logic of science. Henry drafted naturalist Spencer Baird to lead the new US National Museum. Baird identified museums as worthwhile preserves for research in natural science, which still fundamentally relied on physical collection, and he built the museum's collections systematically. His successor and protégé, George Browne Goode, articulated a vision for the museum as a record of human history as well, recasting artifacts of US military prowess and scattered personal effects as the material evidence of national history. Meanwhile, Smithsonian and Bureau of American Ethnology scientists led by curator and ethnologist Otis Mason established disciplinary logics for classification and display, with plants separated from tools and bones.

The Smithsonian Institution's later repudiation of the National Gallery sealed the fate of the museum in the Patent Office—if not a crass sideshow, it was an amateur exercise, the bastard child of the curiosity cabinet, no longer rarified as the province of gentlemen, but amassed by every man or woman with a professed interest in nature. Even its location in an undedicated space seemed to indicate its failure. Yet this presupposes that the Patent Office ought to have been a place where property rights were distributed, not a place of public research or education, and that museums were places where knowledge was shared, not places where property was amassed. The new arrangements of knowledge obscured the fact that these were chiefly the same collections, reshuffled. The National Institute gave the sciences in the United States a material base rooted in naval expeditions, consular networks, and geologic surveys.

Collecting objects removed them from their geographic, material, and social contexts. The specimens in the Patent Office no longer served the purpose for which they were created. Through exchange and display, they acquired new meanings. Chopsticks were not tools for eating but representations of Chinese custom. Opium pipes were not for smoking, but an indictment of Chinese temperaments. The skull of the Fijian chief Ro-Veidovi, taken prisoner by Charles

Wilkes during his expedition in the South Pacific, became not the body of a man or a prisoner of war, but a tableau of Fijian cannibalism, and by extension a proof of American civilization's triumph over savagery.[52] American Indian relics became the symbolic capital of territorial expansion, gathered in tandem with the removal of indigenous Americans west of the Mississippi. Collectors, curators, scientists, and statesmen rationalized these acts of collection as salvaging the material evidence of peoples predestined for extinction. This logic hastened destruction and recast it as preservation.[53] The seeds stewarded by the same people were recast as products of nature, the raw materials for improvement.

The logic of Euro-American property forms constrained the classification of the material collected in, and later deaccessioned from, the Patent Office. Predominant modes of nineteenth-century exchange reserved ownership entitlements for collectors, whether objects were gathered, gifted, or purchased. Only after their initial acquisition were specimens subject to rules of property, including variable and contested customs of ownership and exchange. While custody might change over time, objects retained their significance based on places of origin. In making geography rather than function or singularity the determining factor of an object's significance, collectors framed culture as geographically particular and identifiable. Their dual focus on property and place of origin laid the groundwork for future concepts of heritage and cultural property waged in debates over the ownership of intellectual and material resources.

Meanwhile, plants, tools, and bones were refigured a common treasury of nature disjointed, prior to, and proof of human development. Their very mass, which had threatened their value as rarities, became proof of their status as common objects, ubiquitous and shared, the common heritage of humanity. New institutions of scientific agriculture, including land grant colleges, state experiment stations, and the USDA, absorbed the Patent Office's program to collect and distribute global seeds. The technological detritus retained in the halls of the Patent Office Building acted as a monument to manufacturing interests and the grounds for an increasingly facile material narrative of human development.

While the Patent Office attempted to amass, represent, and reconfigure wide-ranging practices of knowledge making through collection and exhibition of the natural and human-made worlds, this effort did not render it a center of calculation or a locus of rationality so much as a single node of circulation for currents of knowledge and material flowing in many directions. The material in the Patent Office Building museum arrived through a long chain of social and commercial transactions. Before and after conveyance to the Patent Office, each was the object of contested claims to technical knowledge and rights of ownership. Beyond the walls of the Patent Office, in workshops and wheat fields and sickrooms, people borrowed, copied, improvised, and created. Like the collections in the National Gallery, everyday claims to useful knowledge were a disordered, disputed affair.

The people who visited the museum had little knowledge of the controversy that produced it, but they experienced its ambitious disorder nonetheless, encountering products from around the world alongside specimens of American manufactures. Buttons from Connecticut shared the hall with necklaces of human teeth from Fiji—both elaborate ornaments of bone to adorn the body. Plowshares from the Pacific had their western counterparts, advanced enough to deconstruct the wild grass prairie of the American Midwest. The value of the museum's collections remained in question, subject to variable logics of market, utility, and rarity.

The seeds and plants themselves proved unstable, subject to multiple definitions, rules of ownership, and claims of access. Rather than resolving the debate over the value of museum specimens, the parties in question parted ways, dividing their possessions with varying degrees of amicability and hostility. As the natural sciences diverged from ethnology from history, and the Smithsonian from the Patent Office from the new US Department of Agriculture, each discipline and institution dictated its own regimes of value, along with supporting technologies of preservation, storage, documentation, and exchange.

Wide ranging collecting practices nevertheless persisted in each domain, retaining unspecified suppositions about the relationship between property, possession, and knowledge. Was a seedling in a vented glass case a specimen for nature study? Or the basis of a new crop? Was it the property of the collector, the people from whom it was collected, or the agriculturalist who acquired and planted it? Was it important as an object of nature, or of human creativity? Did its value hinge on its potential yield, or on the knowledge of its natural properties? These debates would persist as new disciplines and institutions took up the charge of promoting science and the useful arts, providing radically different rationales for preservation of seeds as representations of species, objects of beauty, raw materials for cultivation, or objects of research and improvement.

2 : SEED SHARING IN THE PATENT OFFICE

In 1861, a wartime Congress attempting to consolidate the political power of northern farmers legislated into being the autonomous United States Department of Agriculture; the Morrill Act, establishing western land grant colleges; and the Homestead Act, opening millions of acres of land to independent farmers. Together these reforms, which had been blocked by southern congressmen supporting the extension of slavery, aimed to secure the West for agricultural settlement on the basis of free labor. In 1862, the new USDA absorbed the Agricultural Department of the US Patent Office, which had for the last twenty-five years sponsored a scattered array of publicly funded research and development, including the importation of seeds for gratis distribution to American farmers.[1]

Perhaps ironically, the Patent Office, a temple to private property rights in invention, adopted a model of public research and free circulation of specimens that persisted in the autonomous USDA. Through the machinery of the US Navy and consular service, missionaries, and American citizens abroad, the Patent Office amassed and distributed thousands of new varieties of forage and fiber plants, mulberries, tea, legumes, garden vegetables, and temperate and tropical fruits to interested agriculturalists. The program was controversial. While advocates supported the federal government's strong role in introducing new crop varieties, critics decried it as a partisan, wasteful, and interference with the efforts of individual improvers. Nascent seed companies focused in the horticultural sector regarded their varieties as market commodities, subject to the rules of free commerce and/or the protections of private property. By the 1930s, seed firms successfully lobbied for intellectual property rights to protect their products.[2] Even so, the eventual commodification of seeds in the twentieth-century United States was not so much a novel imbalance of private and public interests as it was the logical outcome of federally supported enterprise regarding seeds as instruments of national growth.

The narrow construction of debates over the seed program according to the interests of American farmers, horticulturalists, and seed companies ulti-

Figure 2.1. This photograph, taken in 1860 to document the construction of the Capitol dome, also captured the Conservatory and ten-acre grounds of the Patent Office propagating garden, from First Street to Third Street between Pennsylvania and Maryland Avenues, SW. Library of Congress, Prints and Photographs Division, LC-USZ62-86303.

mately obscured more fundamental and long-standing inequities in the collection and distribution of global resources. These included the continued appropriation and exploitation of indigenous American cultivated plants and the reliance on European colonial expropriations in Asia, Africa, and the Americas. That is, while the politics of distribution played out fiercely in the antebellum United States, the politics of collection remained obscure.

Although economies of sharing were divisive, they were fundamentally nationalist, regarding global nature as a reservoir to be tapped for national development. And even when seeds were considered objects of common use not subject to property rights in innovation, they remained subject to complex formulations of access and possession. Notions of commons, collectivity, and mutuality allowed sharing for some but not others and effaced the appropriation of global resources and knowledge to support national development.

The Patent Office's seed program thus shaped a culture of public research that denied the global politics of plant collection, laying the foundation for lopsided legal and scientific narratives of agricultural innovation that elevated the claims of researchers over farmers. In casting improved varieties as objects of innovation, agronomists refigured collected material as unimproved, the raw material for subsequent research. This culture of the commons justified an ordering of global resources that generated broad inequities between countries sponsoring capitalized research and those from which they collected, and between institutionalized research and development and the broad field of agrarian knowledge on which it relied. Arguably, these varied claims persist in the administration of public seed banks situated between legacies of imperial exploitation, private enterprise, and public research.

When live plants traveled through circuits of Euro-American capitals, it was in glass vented cases and wood crates packed with straw, the latter of which sometimes arrived waterlogged with dead plants inside. There were numerous problems with shipping, loss, damage, and theft, partly accounting for the high rate of failure with introduced seeds. Seeds and plants thwarted the efforts of improvers by dying, getting sick, attracting pests, and otherwise proving themselves averse to new climates and geographies. Nevertheless, along with innovations in mechanization and fertilizers, transplantation from other geographies was a principal mechanism of American agricultural improvement in the nineteenth century.

Modeling European imperial exploration and practices of natural science, the Patent Office adopted variable practices of collection based on commerce, gift, exchange, and smuggling. This enterprise spanned the Atlantic and Pacific worlds and involved a heady degree of speculation in the ecological and economic prospects of new crops. By the 1850s, an international commercial marketplace of seeds and plants centered in western Europe took shape, supported on one hand by French and British colonial ventures and on the other by technologies of steamships, Ward cases, and elaborate printed catalogs.[3]

While these efforts looked to the preceding two centuries of European practice, there was nothing uniquely modern in the Patent Office's transplantation and improvement of seeds. As it had been for millennia, mass selection remained the primary method of plant breeding. When Europeans settled in North America, they transported with them crops that had come to western Europe from Africa and Asia during the preceding centuries of agriculture and trade. In service of mercantilist political economy, colonists also mined the continent for new and useful plants. European settlers took up Native American food crops, including numerous varieties of maize. As wheat, apple trees, and agricultural weeds from Europe sprouted, ginseng and other native plants became targets of exploitation for international trade.[4] Production supported by the Atlantic slave trade dramatically altered the ecology of Europe and the

Americas after 1500, moving plants, animals, and microbes across oceans on an unprecedented scale.

Cash crops were a special target of enterprise. The native plant of tobacco flourished on the Eastern Seaboard, although it rapidly stripped the soil, requiring long fallow periods or access to uncultivated land.[5] Rice, cultivated by African slaves in the South Carolina low country, satisfied Europeans that the climate and geography could support a range of tropical and Asian plants. Cultivation also exploited the knowledge and labor of African slaves, who conveyed to South Carolina both African varieties of rice and elaborate techniques of irrigation, cultivation, harvesting, and processing acquired over generations.[6]

Early republicans such as Thomas Jefferson imagined an agricultural nation with a cosmopolitan and global orientation. Propertied elites formed agricultural and scientific societies populated with ties to the Jardin des Plantes in Paris, the Royal Society in London, and the new Royal Botanic Gardens at Kew. With his vision of the United States as an agrarian republic, Jefferson dedicated special energy to agricultural improvement. He pursued wheat resistant to the Hessian fly, which had invaded the United States in force by the 1790s. He also imported varieties of rice from China, Italy, Egypt, Palestine, and equatorial Africa.[7] Indigo, cotton, and silk were also targets of experimentation by southern planters. Perhaps the greatest inspiration for improvers was the introduction of Anguilla cotton to the upper South in 1785 by loyalist exiles, creating the preconditions for the cotton boom of the next three decades. As in the case of rice, West Indian and South Carolinian planters relied on their slaves' knowledge of cotton culture to facilitate technology transfer.[8]

The state explicitly supported transplantation ventures, attempting to bring the superior resources of the government to agricultural improvement. In 1819, Secretary of the Treasury William L. Crawford formalized requests that the US consuls and navy assist in the introduction of new seeds and plants. Secretary of the Treasury Richard Rush renewed these petitions in the 1820s, distributing a circular to diplomats soliciting collection. Declining a proposal by the New York Horticultural Society to partner with the federal government, Rush ran importation through his office. In the city paper, Rush invited interested members of Congress to contact him for a portion of wheat and barley from Tangier that might be successfully cultivated in the southern part of the union.[9] If early nationals pursued agricultural and horticultural improvement according to the imperatives and traditions of European maritime commerce, however, their resources paled in comparison to the European infrastructure of colonial and metropolitan botanic gardens.

It was in large part the initiative of an ambitious new commissioner of patents that boosted federal efforts to introduce and distribute new and improved seeds to American farmers. Beginning with Henry Ellsworth's tenure as commissioner of patents in 1835, the leadership of the Patent Office dedicated funding to importation, propagation, free distribution of new seed varieties, and the production and circulation of statistics and agricultural research on soils, fertilizers, and pests. Ellsworth was a noted agricultural expert with an interest in farm statistics, and he framed the program as a counterweight to the patent system's emphasis on limited monopolies. Yet Ellsworth's politics were more significant for their generally expansionist orientation than for their antimonopoly sentiment, and it was this aspect of the Patent Office's seed distribution that persisted even after Ellsworth left his post in 1845.

Ellsworth claimed agricultural inventors inspired the program by bringing local seed varieties of garden vegetables and maize with them when they visited the Patent Office on other affairs. By encouraging their donations and providing space and infrastructure for exchange networks, Ellsworth carved out a space for common use in a temple of private property. He envisioned the Patent Office as a "clearinghouse for a national friendly community of seed sharers," as Philip Pauly has put it, modeled on the tradition of sharing of seeds as a form of mutual aid.[10]

As Pauly remarked, at first blush it may seem ironic that an organization created to issue limited monopolies for technological innovation would collect and distribute self-replicating natural objects free of charge. But these efforts fell within the Patent Office's mandate to promote science and the useful arts. Unlike the numerous machines employed to cultivate them, seeds were excluded from patent protection, nor would patents have been easily enforced for naturally reproducing objects requiring little capital for production.[11] Ellsworth interpreted broadly the agency's mandate to promote science and the useful arts, including the encouragement of agricultural improvement beyond the bounds of the patent system. In addition to the seed distribution program, he sought to make statistical information on meteorology and production freely available rather than mediated by metropolitan financial interests.

The effect was a bifurcated political economy of innovation, carving out a zone for common use within a regime largely dedicated to buttressing private property rights in invention. Ideologically, these initiatives allowed Jacksonian Democrats averse to monopoly to rationalize patents for inventions by providing comparable incentives to farmers whose improvements lacked property protections. Politically, they addressed the single largest voting bloc in a still-agricultural nation.[12]

Yet Ellsworth's redirection of funds to agricultural improvement was as motivated by his interest in the development of western lands as it was his antimonopoly sentiments, and in this he was consistent with the better part of Jacksonian Democrats. Following the Indian Removal Act of 1830, President Jackson had appointed Ellsworth, a lawyer and businessman with land holdings in Indiana and the prairie states, as US commissioner of Indian Tribes in Arkansas and Oklahoma. Charged with settling land disputes in the new territories, Ellsworth traveled west to investigate and resettle tribes according to the newly dictated boundaries. Jackson rewarded him for his efforts by appointing him superintendent of patents in 1835, from which he was elevated to the status of commissioner a year later.[13] In making agricultural improvement his special cause, Ellsworth thus represented his own interests not only as an agriculturalist, but also as an investor in western lands.

Antimonopoly rhetoric concealed the extent to which western expansion relied on older styles of European imperial exploration in support of national economies. When Ellsworth lobbied Congress for the dedicated funding in 1839, it was to administer the collections of the US Exploring Expedition to the Pacific. The Pacific expedition was meant to announce the arrival of American science on a world stage, and the service of science to the state was well known.

In characterizing the US Exploring Expedition as a boon for American agriculture, Ellsworth proved himself fluent with traditions of European maritime commerce and the networks of naturalists they supported. In his appeal, he noted the failures of the navy's unsystematic efforts in plant introduction to date and urged the Patent Office's expanded role in agricultural improvement. He further advocated the exhibition of plants in the Patent Office, and at the cost of sparring with Wilkes and the National Institute for control of the galleries, he succeeded.[14] In 1840, Ellsworth formally took charge of crop introduction, securing an agricultural appropriation for $1,000 to aid in the production of agricultural statistics and the collection and distribution of seeds, which it offered free of charge.

Henceforth the botanical specimens of nineteenth-century government science expeditions doubled as seed banks for American farmers. William D. Brackenridge, a Scottish nurseryman and botanist who had worked for the nurseryman Robert Buist before joining the US Exploring Expedition as a naturalist, was responsible for the first lot: thousands of plant and bird specimens from the Wilkes Expedition spilled out of the cellars and garrets of the Patent Office Building. Their arrival inspired the hasty construction of a greenhouse behind the Patent Office to house the five hundred species of over 1,100 plants in cultivation brought home by the expedition. In spite of congressional restrictions on the circulation of specimens that so troubled the National Institute, Brackenridge continued exchanging duplicates for new species, as did his successor, William Smith. By 1844, a second greenhouse and addition had been constructed behind the Patent Office, with collections augmented by exchange, donations, and the returns of new naval expeditions.[15]

Ellsworth's rhetoric of agrarian mutuality gave the Patent Office programs multiple protocols of exchange. In drawing simultaneously on networks of Euro-American natural science and agrarian cooperative association, he invoked traditions of seed sharing with very different assumptions about the status and significance of the materials in which they trafficked. The norms and etiquette of naturalists' exchange networks rendered seeds the property of the recipient, who could preserve, exchange, or liquidate collections according to his or her own judgment. Models of seed sharing as mutual aid, in contrast, cast seeds as a common property, implying an ethical burden to a wider community of farmers. Meanwhile, in legal doctrine the common construction of seeds as products of nature denied the human labor of selecting and stewarding seeds, rendering them inadmissible for patents and implicitly sanctioning collection without regard for rights of ownership.

In each of these models, the noncommodification of seeds raised questions of access and custody, rendering them subject to shifting claims of entitlement.[16]

Although the seed distribution program flagged after Ellsworth's tenure ended, the basic infrastructure he established remained in place, including its overt orientation toward western expansion. Ellsworth was sufficiently agile and competent to remain commissioner through the administrations of Andrew Jackson, Martin Van Buren, William Henry Harrison, and John Tyler, eventually ceding his position when Democrat James K. Polk took the presidency by pledging to annex Texas.[17] Between 1845 and 1862, when the newly formed US Department of Agriculture absorbed the Agriculture Department, the US Patent Office continued to enlist diplomatic, missionary, military, and commercial agents to support the expansion and diversification of American plant resources. They included new varieties of wheat, corn, cotton, and a wide variety of Mediterranean and East Asian plants deemed likely to flourish in the climatically similar American South.

In 1849, the construction of a new wing of the Patent Office Building required the relocation of the structures to a site near the Capitol building, formerly home to a botanic garden managed by the Columbian Institute before its dissolution in 1838. In 1857, a propagating garden was established nearby, where promising transplants, including more than 50,000 tea plants from China, were raised for their eventual distribution to American farmers.[18]

The persistent efforts of the Patent Office relied on the consular service, private citizens, naval expeditions, and commercial channels. While they were minor figures and patronage appointments largely ignored by the secretary of state, from the Revolution onward consuls provided a stable network of US presence in the world, often working in concert with the US Navy, independent merchants, and missionary societies.[19] Private citizens in commercial and missionary capacities provided another prolific source of seeds and plants for the Patent Office's seed-sharing program, often working through the consular offices. Especially sustained attention was devoted to South American agriculture, generally calling on a comparatively well-developed diplomatic and consular network.[20]

Government-sponsored scientific expeditions provided a wide-ranging and systematic source of new seeds from South America, the Mediterranean, and East Asia. The Wilkes Expedition became the model for subsequent voyages in the 1840s and '50s, including the Herndon-Gibbon Expedition to the Amazon, the Naval Astronomical Expedition to the Southern Hemisphere, and the Page Expedition to Rio Paraguay and Rio de la Plata.[21] These imports picked up on earlier interests in South American plants, including the possibility of cultivating cinchona, the principal treatment for malaria and an important object of British colonial agriculture.[22]

Naval expeditions oriented more explicitly toward commerce, too, retained a scientific corps, as with James Morrow's collections under the auspices of the Perry Expedition to Japan. Via James Dobbin, then secretary of the navy, he forwarded specimens of vegetables, barley, rice, beans, persimmon, tangerine, and African wheat to the Patent Office from his stops in Brazil, South Africa, Java, and China en route to Japan. With the aid of the US consul in Singapore, Dobbin also forwarded cotton and sugarcane from Mauritius and Singapore collected from local sugar estates.[23] In his schedule of gifts and other acquisitions of the expedition, Perry also listed numerous Chinese and Japanese plants, ornamental trees, fruits, flowers, and sugarcane cuttings in Morrow's care on the store ship *Lexington*. And he noted that Morrow was bringing with him to the United States a young Chinese gardener schooled in cultivation.[24]

Although individual civilians, missionaries, and consuls reported on many aspects of cultivation, from family cultivation to local use of medicinal plants, the Patent Office chiefly pursued hardier and more high-yielding varieties of staple crops, often reverting to the most reliable and well-known varieties verified by a generation of colonial practice. Sugarcane was one target of improvement. In 1856, Congress approved a $15,000 appropriation for the collection of South American sugarcane to aid failing Louisiana planters. William L. Marcy, secretary of war under James K. Polk and secretary of state under Franklin Pierce, forwarded numerous seeds and cuttings gathered by US consuls abroad.[25]

Another target of improvement was wheat, which was especially susceptible to pests, fungus, and chill. The Patent Office cast a wide net in its search for hardy varieties resistant to the Hessian fly, locusts, rust, and cold temperatures. In the 1850s, samples came through the offices of consuls and the US Navy from Spain, Iona, Poland, Turkey, Syria, Algeria, and Chile.[26] The Patent Office also looked for prime specimens closer to home. Gustavus de Neveu of Fond du Lac, Wisconsin, was asked to send his best-quality spring wheat. He forwarded three barrels (1.5 bushels) of Rio Grande Spring and Canada club wheat.[27] I. W. Buchanan, a farmer in Tullahoma, Tennessee, sent a specimen of red wheat known as the "Walker wheat," writing that Tennessee farmers considered it the very best of the red variety.[28]

The Patent Office relied on a vast network of volunteer labor and interested farmers, who wrote to the commissioner donating seed, requesting new varieties, and offering reports on experimental trials. The Patent Office distributed anywhere from one-half bushel to twenty sacks of grain to agricultural societies[29] and individual farmers, many of whom reported on their experiments with new varieties.[30] Among these was Isaac Newton of the Philadelphia Agricultural Society, who would serve as the first commissioner of the USDA from 1862 to 1867. In October 1852, Newton requested half a bushel of Mediterranean and Chilean wheat for a trial.[31] Farther west, agricultural societies supplied native seeds in exchange for the Patent Office's samples. In July 1855, the Fenton Agricultural Society in Northwood, Minnesota, responded to a letter from the Patent Office, reporting on varieties of wheat and vegetable seeds, acknowledging receipt of Turkish and Iona wheat, and stating that it would forward wild rice procured from the Indians.[32] Others reported on trials and made requests for additional seed: in July of 1856, John Henry of Mount Erin, Indiana, reported favorably on Turkish white flint wheat and requested Wyandot corn.[33]

Volunteer labor is notoriously unreliable, and some contributions were more notable than others. Grains of wheat turned up in mummies on a semiregular basis, perhaps linked to the onslaught of European and American Egyptomania in the mid-nineteenth century, which included a fashion for public unwrappings. In 1854, Daniel Somers of Ravenswood, Virginia, submitted to the agricultural department of the Patent Office a package of seed "said to be the kind that Joseph's brethren went to Egypt for." The seed, he noted, was alleged to be the produce of one grain found in a mummy over 2,000 years old. His submission was logged, and Somers received samples of Oregon peas and rice with gratitude.[34] Nine months later, John Reed of Huntington, Pennsylvania, also sent "Egyptian mummy wheat"; whether this was from the same mummy or a different one is not clear.[35]

These were not the dreams of fools; or at least, Ohio Secretary of Agriculture John Hancock Klippart echoed them in his 1860 treatise on the wheat plant. "It is well known to every one conversant with the history of Egypt, that the culture of wheat there has long since been abandoned, and no wild plant in any respect resembling the wheat plant is found," he observed. "But from engravings on ancient tombs at Thebes of the details of plowing, sowing, harvesting and garnering this grain there is no good reason to suppose it has not been cultivated in Egypt from the earliest of this nation's civilization." He also noted the presence of wheat seeds as grave goods: "In the sarcophagi of many of the Egyptian kinds or nobles, were found in vessels perfectly closed, good specimens of common wheat, so perfect kneed that not only the form, but even the color was not impaired, although it must have been inclosed [*sic*] for many thousands of years." Klippart repeated the alleged histories of mummy wheat's provenance without comment.[36]

Still other fantasies of salvage were linked to national patrimony. The postmaster in Tivoli, New York, submitted a small box of wheat he had buried underground near the Hudson River before a British fleet in 1777 burned a stone house filled with the same.[37] W. Noland, the former commissioner of public buildings, forwarded the produce of grains cultivated by his grandson, which he alleged were recovered from a case containing a statue of George Washington many years since. "When the case, containing the statue of Washington was opened in the rotunda of the Capitol," he explained, "there was discovered in the straw, with which it was packed, a few grains of wheat of a superior quality."[38] These "curious specimens of 'Italian Wheat'" seem to have been distributed to a number of recipients, at least one of whom returned some of the product to the Patent Office. The Patent Office redistributed them to agricultural societies in the wheat-growing states. At the suggestion of the donor, a sample was also sent to Mount Vernon.[39]

These claims to patrimony echoed the efforts of early republican naturalists to locate species of grain endemic to North America. DeWitt Clinton, New York senator and avid naturalist, earned the approval of the Linnean Society of London for his specimens of wild wheat in Oneida County, which he thought evidence of North America's claim to rival West Asia as a cradle of civilization. He had accomplished the same feat for rice, locating a species growing wild in the Montezuma swamps of the Seneca River that formed a primary part of native diets. The Fenton Agricultural Society's contribution of wild rice indicated continued interest in wild rice as a staple grain.[40]

But by midcentury, many of these researches had ceded to romances of historical archaeology. Commissioner of Patents Charles Mason was inclined to dispel some of the mystique surrounding salvage seed. "A grain of wheat is found in the crop of a wild goose, another in a chest of tea from China, and a third by accident vegetates in a cleft in the rocks, which shoot up alone into a vigorous growth," he expounded. "These become respectively the progenitors of the Goose wheat, the Tea wheat, and the Rock wheat. For a few years, each acquires a great reputation in the agricultural world, and then relapses into mediocrity. What is the explanation for these phenomena? Why, simply that each of these grains of wheat was originally nothing very extraordinary; . . . When sown broadcast and left untended . . . it relapses into its original condition."[41] There were no miracles.

Yet even the Patent Office's transplantation efforts were a historical romance of agricultural prosperity. Although skeptical of miracle seeds, Mason recommended continued efforts to discover varieties. In the summer of 1854, he charged the US legate at Constantinople, John P. Brown, with procuring one hundred bushels of winter flint wheat from near Mount Olympus or Mount Iola.[42] Sensing the importance of pedigree in matters of breeding, Brown also wrote to Palestine for one hundred bushels of wheat of Abraham's Farm at the foot of Mount Carmel.[43] Plant exploration persisted on the basis of geographic and archaeological fantasies, a legacy that would endure in the succeeding century.

At least outwardly, commissioners in search of useful plants claimed to serve the interests of both eastern farmers battling depleted soil and western ones in custody of newly tilled land, but this expansive address concealed tensions between advocates of free labor and slavery in the North and South. As lands annexed during the Mexican War dramatically increased the southwestern frontier, the Whig Party disintegrated around the question of whether territories would permit slavery.

Conflicts over slavery erupted in the halls of the Patent Office, the agricultural programs of which might sponsor either free labor or plantation cultivation. In 1849, Commissioner of Patents Thomas Ewbank's strong antislavery convictions precipitated the removal of many proslavery staff, including

Southern Cultivator and *Genesee Farmer* editor Daniel Lee, who had managed the agricultural division in part based on his expertise in cotton culture. Almost as soon as he was hired, Lee complained to the secretary of the interior that Ewbank had not paid his salary. Ewbank in turn never acknowledged Lee's employment by the Patent Office. When Taylor's death elevated Millard Fillmore to the presidency in 1850, Ewbank, a supporter of Fillmore's archrival and abolitionist William Seward, was obliged to resign.[44]

Ewbank's replacement, Charles Mason, towed a more cautious line politically. Acquiescing to Fillmore's general strategy of appeasement, he framed the Patent Office's agricultural work more broadly than had the proslavery apologist Daniel Lee, who complained that his opponents simply resented his statistical demonstration of the overall productivity of the slave system. In 1853, the newly appointed Mason summarized the department's charge as the promotion of plant introduction and improvement, soil analysis, and agricultural expansion. The last of these encompassed a wide range of projects addressed to different constituencies, including "the interests of the farmers and planters of the United States in the improvement of their crops and live stock; the introduction of new and valuable products; the amelioration of exhausted and unimproved soils of the States lying along the seaboard and the Mexican gulf; and developing the agricultural resources of those bordering on the Pacific, the Mississippi and its tributaries, the Great Lakes, and the Canada frontier, thereby producing larger quantities and of better quality, of our chief staples for export and domestic use."[45] Mason's capacious formulation obscured the intensifying focus on the political economy of the new western territories, and specifically whether a plantation system reliant on slave labor would flourish there.

The Patent Office's rangy and noncommittal politics with regard to the future of slavery inspired renewed efforts to locate and transplant promising cash crops for American farmers in the new southwestern territories. In his effort to indirectly address the overproduction of cotton in the South, or to dodge the subject entirely, Mason renewed earlier petitions to US consuls to gather useful plants. Mason and his agricultural clerk, Daniel J. Browne, skirted conflict by suggesting the range of possible crops that could flourish in the West, while asserting the viability of cotton in the same regions. Emphasizing the similarities between climates in East Asia and the American South, the Patent Office's energies in the 1850s ranged widely to include the opium poppy, soy, and sorghum, among others. While speculation in silk had fizzled by the mid-1840s, hampered by extensive labor requirements for cultivation and inflated prices for seedlings at the height of the craze, Browne was undeterred by the previous decade's speculative failures. He recommended tea as the next cash crop that could rival cotton in the American South, building on a series of failed introduction efforts to date.[46]

In spite of the overt controversy, the activities of the Agricultural Department did not clearly express any one partisan agenda or economic interest, but rather reflected a broad political commitment to national expansion. The seed distribution program developed in tandem with military actions to annex western lands. Ellsworth's inauguration of the seed distribution program took place on the heels of the Indian Removal Act. Mason revived it in the wake of the Mexican War. The Patent Office's effort to import and distribute seeds and cuttings for the benefit of American farmers was part of a concerted exercise of military force to secure the American West for Euro-American agricultural settlement.

The seed distribution program faltered politically not on the issue of territorial expansion or the proper labor model for agriculture, but rather on competition with the private sector. Perhaps the most problematic source of seeds for the Patent Office was also the most secure: European botanic gardens and commercial nurseries. Overreliance on these channels made the program susceptible to charges of waste and challenges from American seed dealers, who saw competition rather than aid. Controversies over seed distribution expressed themselves in partisan terms, but these masked a murkier relation between government-sponsored science, private networks of exchange, and the many different commercial producers of seeds competing in the antebellum marketplace.

While the Patent Office pursued extensive transplantation through naval exploration, it established seed exchange programs with European botanic gardens in Paris, Zurich, Berlin, Leipzig, and Baden. Through the intervention of Alexander Vattemare, a "well known and philanthropic gentlemen," the minister of agriculture and commerce in France forwarded numerous specimens cultivated in the Jardin des Plantes prepared by the professors and curators of the Museum of Natural History in Paris. He also forwarded cases of specimens from the Algerian provinces, collected at the order of the minister of war to be presented to the Patent Office. Vattemare was a major proponent of exchange, which he interpreted as a means of securing specimens not otherwise obtainable, in this case enclosing seeds conveyed by the minister of war from the Algerian Annual Agricultural Exhibition.[47]

Often these formal arrangements did not function as truly free systems of exchange, especially as the Patent Office became susceptible to charges that too many seeds came from predictable European commercial channels. C. P.

Hagedorn, the consul in Bavaria, was a tireless correspondent, establishing connections with botanic gardens in Leipzig and Baden. By 1856, the Patent Office noted that the last three years' shipments had come from standard commercial channels and asked that subsequent samples be of Bavarian origin. When Hagedorn requested samples of American tobacco wanted by the Bavarian government, the Patent Office referred him to B. L. Jackson & Brothers of the Pennsylvania Society for Seed, indicating the relative maturity of commercial seed firms in the United States and the Patent Office's willingness to work through them.[48]

In practice, the boundary between state botanic gardens and commercial nurseries was blurred, and the Patent Office might exchange seeds with nurseries in addition to placing orders on credit. Seeds donated by the French seed dealer Vilmorin Andrieux and Co., for example, were placed in the collection of the Patent Office museum with the commissioner's gratitude. In exchange, Vilmorin requested seeds and cuttings the Patent Office deemed appropriate, including the "Sequoia giantea, the Thuya gigantea, and in general all of the coniferous trees of California and Oregon."[49] These were showpieces rather than raw materials for cultivation, but they were exchanged through the same networks as garden vegetables and seed grain. The porous boundary between public and private did not trouble the leadership of the Patent Office, but it would become a problem for domestic nurserymen and improvers who felt excluded from the government trade.

Gradually, a reliance on established European seed dealers with colonially forged networks of collection displaced more decentralized private seed exchange efforts as well as more ambitious attempts to collect through naval expeditions and consuls. In 1853, Mason sent Browne to Europe with the task of procuring seed varieties, primarily from French commercial dealers. Vilmorin in Paris and Charlwood & Cummins in London were the Patent Office's principal sources of seeds after 1856, even for its most ambitious far Asian transplants. In 1860, Vilmorin supplied the Patent Office with Mediterranean wheat and Chinese sugarcane, whereas six years earlier it paged US consuls for such acquisitions.[50]

The Patent Office also established commercial relationships with domestic seed dealers, including Bissell for flower seeds, Comstock Ferre and Co., J. M. Thorburn, and Prince's Nursery in Queens, the latter of which advised the Department of the Interior on the possible formation of a national nursery.[51] Critics charged that the seed distribution program had become a traffic in commercially available varieties meant to secure votes for congressmen and favor from the select seed companies tapped to supply the goods.[52]

In this climate, charges of partisanship abounded. Congressmen and postmasters controlled the bulk of distribution of seeds and reports, with the former liable to favor their constituents and the latter reputed to be political appointments, more oriented toward newspaper editors than working farmers. The Ohio Democrat Samuel S. Cox charged that his "black republican antagonist" flooded constituents with free seeds. A Republican claimed that the postmaster distributed all the reports to Democrats. Still another concluded that until these biases were corrected the whole initiative was a great humbug.[53]

When partisanship wasn't a problem, elitism was. In one circular, the Patent Office invited postmasters to nominate agriculturalists in their districts who would be worthy of receiving sample seeds. Established agricultural societies were the most likely candidates, but this practice too invited charges of favoritism, with the farmers most deserving of assistance allegedly bypassed for gentlemen whose wives tended flower gardens. Even when the object of culture was cotton or corn rather than ornamental flowers, elite planters and agriculturalists rather than common farmers received seeds. The reforming southern planter Edmund Ruffin, for example, was a frequent correspondent of the Patent Office and a recipient of seeds from both the Wilkes and the Perry Expeditions, among others. He also contributed to the Patent Office reports on occasion, in spite of developing an animus against Daniel J. Browne so intense he made it his personal project to see the clerk expelled from his post. Toward that end, he published multiple opinion pieces in the agricultural papers, reflecting in his journal that he had freely expressed his "contemptuous opinion" of the man and "the whole working of his department."[54]

Ruffin, an apologist for slavery who fired the first shots at Fort Sumter, may have reviled Browne as much for his antislavery sentiments as his cronyism. He charged Browne with stacking his department with incompetent tools rather than capable farmers, enlisting enough support from the congressional advisory board on agriculture to order an audit of the department. Ruffin and others charged Browne with lacking scientific credentials, committing plagiarism, and wasting Patent Office funds on European boondoggles for common seeds.

Although the final review was mostly favorable, the committee did advise more prudence in determining which seeds were valuable prospects for introduction. It also advised that the Patent Office's nascent experiments with tea ought to be seen through to a successful end, a reflection the expenditures for it, its perceived promise, and the office's history of dead ends.[55] In spite of being cleared of wrongdoing, Browne resigned in short order.

Ultimately, the many constituencies making demands on the Patent Office disagreed as to whose interests its agricultural programs should serve. Exotic and ornamental plants, staple grains, and potential cash crops appealed to different communities, each of which appealed to the Patent Office for aid. Ellsworth and many others may have hoped the diversification of southern agriculture would break its reliance on cotton, but as the political landscape of the office changed from administration to administration, as much attention was devoted to collecting agricultural statistics on cotton's cultivation and production as to locating promising new crops. Meanwhile, the Patent Office's primary chemist, Charles T. Jackson of Boston, devoted a year of labor to analyzing the specimens of Sea Island planters.

But statistics, like chemical analysis, could serve many masters. The Yankee ethnologist Lewis Henry Morgan implored the leadership of the Patent Office to resume its initial attempts to collect agricultural figures by state. Perhaps statistics would confirm what many claimed to know already: the South was overproducing cotton. Daniel Lee, in contrast, claimed his statistics showed healthy production in the South. Other southerners, such as H. C. Williams of Texas, were hostile to the whole enterprise, charging the Patent Office with favoring the Yankees in the development of viticulture, alluding to Charles Jackson's experiments measuring the sugar content of grape varietals. Williams, expressing his distaste for newfangled scientific instruments and for the northeastern urban elite, offered that Jackson could go stick his saccharimeter in some worthless northern juice as he pleased.[56]

As millions of packets of garden seeds circulated under the names of congressmen, several northeastern seed companies attacked the Patent Office's seed program as a threat to the much higher level of horticulture in New England. Although the vegetable and flower seed companies of the Northeast had little business with the cash crops of the South, they nevertheless opposed the Patent Office's intrusion into their business. Major nurseries such as Peter Henderson, Parsons and Co., and Ellwanger and Barry enjoyed subsidized postal rates for shipping seeds, and they dealt increasingly with European dealers associated with imperial botanic gardens, such as Veitch and Loddiges in Britain and Vilmorin in France. Unless the Patent Office directly purchased its seeds for distribution, it was an unwelcome competitor. Moreover, defining plant breeding as an economy of free exchange among farmers threatened their very livelihood as market producers.

Some critics of the Patent Office's seed programs asserted proprietary claims to seeds they cultivated. David Landreth, the prominent Philadelphia seed dealer, had at one time aided the office in the collection and distribution of seeds, recommended by Isaac Newton and the Pennsylvania Agricultural Society.[57] By the late 1850s, however, Landreth found grievance with the Patent Office's programs. By his account, he had originated a new variety of turnip, distributed throughout the United States and later in Europe as the White Strap-Leaved Flat Turnip. According to Landreth, an identical variety was being marketed by the Patent Office as imported from England.

Not everyone accepted Landreth's claims to originality. Some agriculturalists claimed Landreth's variety was simply identical to one grown in England and imported by Charlwood & Cummins through the Patent Office.[58] Nevertheless, the Pennsylvania Horticultural Society, citing Landreth's case, registered a more general complaint on behalf of nurserymen and seed growers, arguing that the seed programs brought "the power and purse of the general government in active competition with their industry." Landreth too attacked D. J. Browne, instigating the Committee on Agriculture's review of the Patent Office department. He and Ruffin made unlikely allies.[59]

While Landreth claimed that the Patent Office violated his rights as an originator, seed distribution raised issues of property for farmers engaged in field trials as well. At stake was the division between public versus private ownership of the fruits of research. P. A. Rett, president of the Southern Agricultural Society of Louisiana, concluded that there were two opinions on the subject. In the first view, seeds and plants received from the Patent Office were "the absolute property of those who receive them," with the recipient neither obliged to sell them nor limited in the asking price. In the second view, the recipient was rather "a trustee selected by the Government for the propagation and distribution of their products," having "the same rights in them as they would have if they were grown on a model farm belonging to the United States." In this role, the recipient deserved market value for the seeds to compensate him for the trouble and expense of raising them, but the duty to extend their cultivation was foremost, and there were implied limits on claims to ownership.[60]

Rett's consideration was prescient, for this tension between claims to property and public good remained unresolved in the public-private relations that characterized the research of the Patent Office and its successor agency, the US Department of Agriculture. In none of these discussions, however, was the labor of originators outside the United States acknowledged. When "ornamental" and "exotic" plants reached imperial botanic gardens or commercial dealers from colonial possessions or ports of trade, it was with little more than an indication of difference and rarity. Botanist explorers such as Robert Fortune were celebrated precisely for their ability to remove plants from their native habitats and transport them to European gardens. Obscurity rather than authorship gave them value. Feats of introduction such as Fortune's were to be admired in part because the originators remained shrouded in mystery.

If the Patent Office's reliance on European colonial networks of gardens and nurseries made it susceptible to attacks from northeastern horticulturalists, its susceptibility to a wildcat market of proprietary breeds weakened its credibility as well. As the agency attempted to respond to the demands of regular farmers, it became a consumer of commercial seeds of questionable quality, many of which made grandiose claims for their products related to both yield and heritage. Notably, these too often gained value by obscuring their origins, mingling claims to provenance and property in a new national marketplace of seeds.

One prominent case was that of Wyandot corn, alleged by its proprietor to have been acquired from the Wyandot Indians. An 1856 edition of the *Prairie Farmer* contained advertisements for Colombian Guano, Superior Devon Cattle, Patent Beehives, and "WYANDOT CORN." The Wyandot corn's alleged productivity rivaled Jack's beanstalk: with common cultivation in a single season, Thomas's nine grains became forty-eight mature ears. The next year, twenty-five grains produced 132 ears; and the next, a fourth of an acre became thirty-two bushels and three pecks of shelled corn. A mere grain per hill yielded up to eight stalks a full twelve feet high with double the usual quantity of ears. Soon numerous agricultural magazines touted Wyandot Prolific as "The Great Agricultural Wonder of the Age!" The advertisement in the *Prairie Farmer* read as follows: "This truly singular production was first introduced to the notice of the public in 1853, by Mr. JR Thomas, of Waverly, Illinois, who received nine grains from the Wyandot Indians through a California emigrant, which he planted in a sandy soil."[61]

It was a dubious claim. The Wyandot, formerly known as the Huron, settled near Ontario and the Georgian Bay in the seventeenth century. Following clashes with Iroquois nations to the south, remnants of the confederation fled westward to what became the Ohio territories, where their next adversary proved to be land-hungry settlers and the agents of federal government. A succession of conflicts and coercive land treaties left the community, by then christened the Wendat, Wyandotte, or Wyandot, with little land, until in 1843 the remainder of its members acquiesced to remove west of the Mississippi to Oklahoma.[62] The population was dispersed enough that Thomas might have encountered a member of the Wyandot community in his travels, though the intermediary of the California immigrant makes his story even less plausible.

Soon after the first notices in the *Prairie Farmer*, Thomas began selling seed through an agent in Staten Island. Advertisements also appeared in select southern newspapers, instructing farmers to obtain seed through a representative of the Georgia Railroad Company. In a matter of seasons, Wyandot corn reached northern, southern, eastern, and western climes through a network of sales agents supported by the railroads and the post office.[63]

Almost as soon as the corn was planted, it had detractors. One farmer who lived on the former Wyandot territory in Ohio questioned whether Thomas had fabricated the seed's provenance. With all respect for his former neighbors, he doubted that they had given Thomas any corn, or that theirs was much to talk about in any event. "If they have propagated a new and more valuable kind of corn since their migration Westward, the fact is unknown to me," the man offered, concluding that the corn was "a humbug of the Morus Multicaulis species."[64] Here he referred to disappointed speculations that silk spun from the mulberry tree would be America's next cash crop.

Popular periodicals quickly downplayed the success of the corn relative to other varieties and criticized farmers' thirst for novelty over responsible husbandry: "earnest seeking after the *new*, the *progressive*, is to be admired," opined one agriculturalist, "but not this mad chase after novelties—this spirit of speculation."[65] Other papers with access to samples took issue with the identification of the corn itself, declaring it a soft white southwestern squaw corn such as those favored by the Indians for home consumption. They warned that the corn was likely only of use for stock feeding. Some deemed it flinty, while others pronounced it soft. Some thought it poor quality but salvageable for distilling into whiskey.[66]

Many criticized the exorbitant price of Wyandot and other Prolifics. "Almost every day startling announcements are made of the discovery of some new variety of wheat, or corn, or grass, or fruit, which in point of productiveness, ease of cultivation, peculiar adaptation to almost any soil or climate, has never been equaled," groused one commentator. "Of course the prices asked for these rare commodities are commensurate with their advertised value. . . . The Wyandot Corn, the value of which remains yet to be tested, finds large purchasers at the modest rate of a penny per grain, or about *800 dollars* per bushel."[67]

For its part, the US Patent Office reconsidered plans to supply samples gratis after Thomas could only supply a single bushel at the exorbitant cost of forty dollars. Anticipating a demand for the corn, Commissioner Charles Mason had tried to procure it but found the available quantity scarce and the price unreasonably high. Ultimately, the Patent Office secured a half bushel, but many recipients complained that it did not vegetate.[68]

Wyandot was not the only agricultural wonder of the age. Wyandot and Peabody's Prolific Corn[69] jostled with Boyd's Prolific Cotton and other proprietary breeds in the basement of the Patent Office Building, alongside less ostentatiously labeled varieties forwarded by consuls, missionaries, naturalists, and naval officers. Demand persisted for the next five years, buoyed by the wide distribution of Thomas's circular and his advertisement in national agricultural periodicals.[70] Farmers continued to lobby the Patent Office to procure the seed, often enclosing circulars and advertisements for Wyandot corn with their letters, or presenting requests in tandem with reports on trials of other seeds received from the Patent Office.[71]

Gradually, the Prolifics waned. By 1860, Ohio Secretary of Agriculture John Hancock Klippart called the corn a "curiosity . . . unworthy of culture," noting that it had been grown "in a very few places," and "not favorably received" in those.[72] But the government never quite succeeded in getting out of the seed business.

❧ A postscript: By the 1880s, state boards of agriculture had endeavored to build up an empirical base of knowledge about the many varieties of cereals in the national market. In 1885, an agriculturalist from the New York Agricultural Experiment Station conducting experimental trials procured a number of varieties of Indian corn from the Smithsonian Institution, including "Wyandotte," which he reported "formerly grown by the Indians of Illinois." He compared it to other varieties collected from the Cocopah, Sonora, Zuni, Tuscarora, and Manitoba Indians, as well as blue corn of Indian origin from Canada and another from San Pedro, Mexico.[73] Wyandot corn entered the Smithsonian National Museum's collections alongside many other alleged American Indian varieties, a number of which were noteworthy for their recent commercial vintage.

Given the emphasis of experimenters on objectivity and empiricism, it is surprising that they took the purported origin of Wyandot Prolific Corn for granted. Enshrined in the collections of the national museum, their provenance appeared to the experimentalist as fact, not legend. In classifying varieties for trials, experimenters lumped commercial varieties of recent vintage in with a wide variety of others, often relying on the collections of other institutions and exercising little scrutiny as to the validity or utility of provenance claims, such that the agriculturalist at the New York State Agricultural Experiment Station could report that the Wyandotte corn he procured from the Smithsonian was formerly grown by the Indians of Illinois.

Of course, at several generations remove, JR Thomas's claim to have procured maize from Native Americans had truth to it. Any farmer who planted corn drew from a variety of northern flint and southern dent corn varieties, all of which derived from Native American maize germplasm. What is more important than the probability that Thomas's provenance story was hokum is that he could only assert it because the alleged progenitors had been forcibly removed from the area ten years before, with no one but a skeptical neighbor left to contradict his tale. While the office continued to investigate American Indian landraces, including wild rice and rope bear grass (a kind of hemp), native cultivators had no voice in the halls of the Patent Office.[74] Instead, they were most visible in the galleries of the museum on the upper floor. Display cases showcased trade goods and weapons collected during the Creek and Seminole Wars, protracted conflicts expressing federal commitment to removing Indians from land coveted for agricultural settlement. Ringing the gallery overhead were portraits of Indian leaders by the painter Charles Bird Kind, who painted diplomats when they traveled to Washington to negotiate the terms of treaties with the federal government.

Thirty years later, the varieties of maize lodged in the Smithsonian, whether genuine or spurious, appeared as the common stock of the American agricultural past and the raw materials for systematic improvement. These designations erased the centuries of labor that had produced them. As proprietors made seeds into commodities through the application of property claims, researchers stripped them of prior human history. Indeed, property claims were often possible only because seeds had been delivered tabula rasa, products of global nature rather than human industry.

Persistent efforts to introduce new plant genetic material were one aspect of the rise of the United States as a global agricultural power, and the new varieties of cotton, sugar, and maize introduced in North America benefited farmers of these staples. New varieties facilitated the consolidation of regional agricultural economies of feed grain, livestock, and wheat in the West, dairy and vegetable production in the Northeast, and cotton and sugarcane in the South.

The Patent Office's efforts to introduce new crops, in contrast, largely failed. Most seeds failed to germinate. Most farmers didn't get them. There was never an institutional commitment to individual small-scale agriculture or extra-market subsistence farming, and the program fell prey to charges of partisanship and waste. Yet the material transfers of seeds and plants were perhaps less important than the precedent they established for agricultural development based on federally subsidized research, exploration, and transplantation. The establishment of these practices made for a seamless transition to a more robust and well-funded US Department of Agriculture, setting US agriculture, for a time, on the path of public research.

However highly politicized the debate over the Patent Office's public research and gratis distribution of seeds, there was a basic consensus that it admitted only political actors in the US sphere, from nursery proprietors to cotton planters, and that it excluded Native American progenitors. Partisan disputes masked the broader geopolitics of collection and transplantation on which the Patent Office relied. Had both proslavery and free labor apologists acknowledged the extent to which continental expansion followed European colonial models, they might have qualified their claims to liberty and novelty.

Ideologies of agrarian mutuality and seed sharing also obscured original appropriations of material through military and civilian networks. The construction of nature as a common resource allowed both British and American explorers to carve out zones of shared resources not subject to rules of trade, market, or variable customs of patronage and reciprocity. As printed texts and plantation labor systems rationalized the skilled labor required for cultivation, plants were cast as products of nature rather than artifacts of accumulated knowledge and technological practice. By removing resources from their points of origin, plant explorers stripped them of their human histories, rendering non-Western and indigenous progenitors invisible as sources of technical knowledge.

The basic research initiated by the Patent Office became the province of the US Department of Agriculture, land grant colleges, and agricultural experiment stations, while applied research became controlled by a burgeoning seed industry intent on expanding markets for their wares.[75] Plant introduction in the US Department of Agriculture continued, supported by the passage of the Hatch Act in 1887, establishing state experiment stations. Although the pursuit of new varieties from beyond European colonial and commercial networks had flagged in the 1870s and 1880s, beginning in the 1890s, Agricultural Commissioner James Wilson sponsored extensive research in Eurasia through the Bureau of Plant Industry and the Section of Systematic Seed and Plant Introduction. Niels Hansen went to Siberia and Turkestan in search of grasses suited to the dry summers and cold winters of the Great Plains. Observing the success of Russian immigrants in the same region, Mark Alfred Carleton searched Russia for new varieties of hard red winter wheat. Once returned, these plants became the objects of federally funded research in land grant colleges and state experiment stations. Carleton's research into durum wheat earned him fame as an introducer and innovator.[76]

While sometimes at loggerheads, ultimately plant breeders benefited from public research. The government acquired plants and distributed them to farmers who worked to make them profitable. The government also continued to sponsor extensive research and development of improved varieties. While they benefited from decades of public research, efforts to breed proprietary hybrids accelerated the commodification of seeds in a private marketplace, ultimately paving the way for the extension of plant breeders' rights and patents on seeds.[77]

Although opponents of corporate consolidation of agriculture identify the commodity form as the primary mechanism of control over nature, as significant were the practices of improvement rendered invisible by explicit noncommodification of seeds. For in spite of the language of collecting naturalists, seeds had never been products of nature; they were always artifacts of human labor, however temporally removed from the nineteenth-century Americans trying to make a good crop. Applying proprietary claims to seeds required that they be stripped of competing claims. When seeds became objects of public research, available information regarding provenance and stewardship rarely did justice to the history of the seed, which was often reduced instead to a point of geographic origin or a vague reference to native cultivators.

To contemporary sensibilities, it is ironic that an attempt to construct a zone of common use within a private property regime in fact generated deeper and more structural exclusions and inequities, as Asia and the Global South provided germplasm for Europe and North America. The imagination of a single world constituted of biologically diverse resources was less a precondition than a product of these efforts, enabling fictions of the global persistently deployed in contemporary contexts.

3 : FAILURES OF TEA CULTIVATION IN THE AMERICAN SOUTH

Down the street from the Patent Office Building was another, less hallowed facility for seed storage, full of seeds no one wanted: the Minor Letters Room of the Dead Letter Office in Washington. In 1852, a roving *Harper's* journalist toured the Dead Letter Office. There, he wrote, clerks labored in "tomb-like" quarters, sorting misdirected love tokens or messages from dead soldiers that languished for months before being consigned to the paper mill. He relished the variety of misdirected parcels in the Minor Letters Room: "fire and life assurance policies, a bunch of keys, a specimen of wheat, bottles, sugar samples, hanks of yarn, a bed quilt, a rattlesnake skin, two diamond ornaments, an old hat, a draft for ten thousand dollars, a paving stone, a suit of boy's clothing, a box of tea nuts from that indefatigable gentleman, Jn. Junius Smith, addressed to some delinquent correspondent, who has omitted to claim them, a pot of ointment, a bundle of watchmaker's tools, maple sugar, a bullock's horn, a galvanic battery, garden seeds, lawyer's papers without end."[1]

It's noteworthy just how many seeds were in the Dead Letter Office; in this one brief list, second only in quantity to the lawyers' papers without end, were specimens of wheat, garden seeds, and a box of tea nuts. In a primarily agricultural nation, the quantity of misdirected seeds addressed to American farmers represented the cost of doing business and the ambitious projects of improvement it supported. The Dead Letter Office was an unlikely catalog of American economic activity, but for the fact that the mails were the primary conduit of business activity in the nineteenth century. The willy-nilly growth of the nation that had made its postal system a logistical marvel also required the establishment of a hub to process misdirected letters and parcels. When they proved undeliverable, they were sent to the central post office in Washington. The contents of parcels, not subject to the protections of first-class letters, were put on display before being auctioned off en masse, becoming objects of curiosity for tourists visiting the nearby galleries in the Patent Office Building. A node of documentation for the articles that lost their way, the Dead Letter Office also provided a record of nineteenth-century American economic activity.

The journalist referred to the sender of one item by name, implying the reading public knew him by reputation. "That indefatigable gentleman," Junius Smith, was an intrepid improver who had spent the previous decade advocating transatlantic navigation by steamship, only recently turning his attention to tea, which only several years before had been smuggled overland from China to British India with great success. Tea, wagered Smith, could unseat King Cotton as the major cash crop of the American South. He modeled his efforts on the success of the British plantation system and believed cultivation in the American South could proceed on the basis of free labor rather than slavery, aided by mechanization.

He failed. By the time Smith's seeds turned up in the Dead Letter Office, he had already been dead several months, clubbed in the head by intruders on his South Carolina plantation the previous year and deteriorating gradually thereafter. His untimely passing was unrelated to the delinquency of his correspondent, however, who had failed to collect the seeds after being supplied free of charge by the US Patent Office's Agricultural Department. The Patent Office too had recently begun experiments in the cultivation of tea. Yet it failed just as Smith had.

Ultimately, there were many failures of tea cultivation in the American South, in Smith's endeavors and those of the Patent Office to unseat King Cotton. These were not simply the failure to acquire live seed or to understand environmental conditions of cultivation, or even to perfect elaborate techniques of production that could produce not one but many kinds of tea. Rather, in imagining tea as a uniform commodity for mass markets, improvers also failed to envision smallholder production with skilled labor as an alternative to the plantation system, and to link their visions for imported seeds to the systems of labor required to cultivate and maintain them. Improvers regarded the tea nut as Jack's magic bean, rather than a materialization of complex systems of labor and knowledge, and as a result, their attempts to develop rural America around its cultivation failed.

The United States, venturing to take its place on the global stage, modeled its transplantation efforts on European examples but failed to achieve their results. By the eighteenth century, European nations that had prioritized expanded trade with China gradually shifted toward a strategy of import substitution, transplanting valuable crops to their colonial possessions. These efforts entailed multiple innovations that shaped global commodity cultures, from new technologies of production and agricultural science to plantation slavery. Because of its comparatively early independence from Britain, the settler colony of the United States followed a different path than British colonies in Asia, characterized by a commingling of private enterprise and government incentives. Its production tobacco, rice, and cotton for export markets nevertheless situated it within an Atlantic world economy.

While systems of production for early colonial cash crops depended on the knowledge and materials of Eurasian merchants and planters, European settlers, and African slaves, by the mid-nineteenth century the United States modeled its efforts on the empire from which it had disengaged. Looking to replicate British success in South Asia, its fledgling state was in some sense already ersatz, readily translating British failures to a North American context. Americans looked on British exercises of botanical imperialism with envy, reporting in wonder on the horde of specimens gathered at the botanical gardens at Kew. Following the British example, they worked to reduce complex cultures of production and consumption to the imperatives of cash crop production for mass markets through plantation agriculture. Yet whereas the British committed extensive resources toward the operation of a ruthless plantation system in Assam, American commitment to transplantation was more scattered and perfunctory, relying on the commitment of planters whose privately held land might already be dedicated to cash crops, principally cotton.

Until the 1840s, the only tea known to westerners came from China. Popular magazines referred to the tea plant as an artifact of mystery. In 1843, mere months after the British military victory reopened Chinese ports, the *Ladies' Companion and Literary Expositor* published a feature, "The Culture and Preparation of Tea."[2] "There has always been an impenetrable mystery enveloping the history and character of the tea-plant," the author began. Knowledge of its botany and culture was spotty enough that most still contended that black and green tea were two separate species of plant. If East Asia had a biological monopoly on cultivation, popular features on tea plants indicated that changes were afoot.

By the 1840s, it had been over two centuries since the Dutch introduced tea from China to Europe, and nearly as long since it had become a fashionable beverage for upper- and middle-class England. It had already played its infamous part in the American Revolution, with the British East India Company's cargo chucked into Boston Harbor. More recently, the East India Company had lost its monopoly on the importation of tea from China. Moreover, it resented China's unwillingness to trade in anything but silver. While traffic in opium provided a destructive wedge into Chinese markets preserved by English gunboats, beginning in the late 1830s the British followed a second strategy of import substitution through the cultivation of tea in British India.

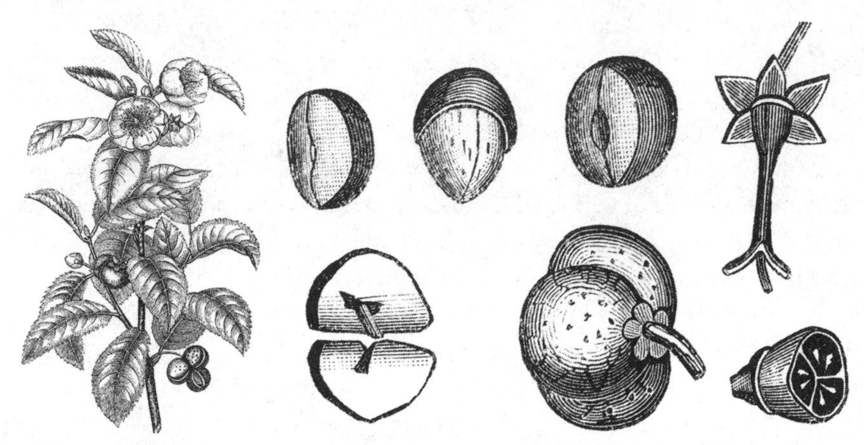

Figure 3.1. Tea plant and seeds, pictured in *Harper's New Monthly Magazine*, November 1859, in an article called "Tea Culture in the United States."

Tea (*Camellia sinensis*) grows throughout China and flourishes in the southern hill country. Although there are multiple subspecies used in the cultivation of green and black teas, methods of processing are the primary factors distinguishing the two types. Black teas are wilted and oxidized, whereas green teas are not. Green teas dominated Chinese markets. Black tea, in contrast, was an export commodity for western European and North American markets. Chinese producers and consumers generally regarded black teas as crude and inferior.

For centuries prior to their transplantation to European colonies, tea, along with cotton and silk, had important diplomatic, economic, and cultural functions in China. Before the European maritime exploration in the sixteenth century expanded China's oceanic trade, the Chinese imperial government attempted to tax and control the overland trade of tea with Central Asia.[3] By the eighteenth century, Canton was the seaport for European trade, with the uplands of northern Fujian supplying nearly all tea for export.

During the eighteenth century, the tea trade with European countries boomed, constituting 70 percent of the Dutch East India Company's purchases and a quarter of the profits of the British East India Company. Between 1730 and 1790, imports of tea to England increased from one to twenty million pounds per year, a source of profit to the East India Company and tax revenue to the crown. Fujian production surged to meet European demand, with total export share increasing from 48 to 73 percent between 1786 and 1830, with silver and opium constituting the principal means of payment. As Anglo-Chinese tensions over the opium trade mounted, the English waged successive military campaigns to maintain access to Chinese ports from 1839 to 1842: the first of the Opium Wars through which the English secured expanded trade rights, including the legalization of the opium trade.

Meanwhile, the British devoted new energy to breaking China's monopoly on tea. British victories in the first Opium War rendered China visible and accessible as a frontier of new botanic resources, and the tea plant was among the most prized. In 1834, Kew botanists in Calcutta had verified that a plant found growing wild in the forests of Assam was a variety of tea, galvanizing hopes that the colony might become a locus of tea production for the mother country. Notably, Kew botanists rejected Assamese specimens for a full decade. Only in 1836, twelve years after Scottish explorer Robert Bruce had first acquired specimens of tea from Singpho chief Bessa Guam, did the British Tea Committee recognize native samples as belonging to the tea family. Even so, British botanists and administrators regarded the Assamese tea plant as inferior and wild, convinced that successful cultivation required Chinese plants and specialists.[4]

Acknowledging its perilous market position, the East India Company partnered with former gunboat captain C. A. Bruce and the governor general of India, Lord William Bentinck, to smuggle tea plants to nurseries supervised by the Royal Botanical Gardens. The company enlisted the Scottish botanic explorer Robert Fortune to locate and smuggle seedlings based on his success in conveying Chinese trees, shrubs, and ornamental plants on behalf of the London Horticultural Society. Fortune's success, building on established networks of diplomats and missionaries, brought thousands of seedlings and a cadre of Chinese experts overland to Kew's outpost in Calcutta and the new plantations of Upper Assam, quite literally supplying the seed of the Indian tea industry.[5]

All this attracted the attention of American agriculturalists, who had flirted with the possibility of growing tea in the United States for years, noting the similarities between the climates of the American South and East Asia and claiming the crop could flourish for local use, home market, and export.[6] By the 1840s, new clipper ships fed the American demand for tea and facilitated opium smuggling up and down the Chinese coast. But American entrepreneurs, like British ones, considered strategies of import substitution. Tea, it was said, could be cultivated on family farms, for the domestic market, and as an export crop to rival cotton and wheat. As a target of speculation, tea took its place alongside silk, another product in which China enjoyed a large market share. Speculation in silk had bottomed out by the mid-1840s, compromised by the escalating price for seedlings, the high mortality of young plants, the long period of maturation, and the labor-intensive nature of cultivation.[7] Agriculturalists also expressed hopes for other Chinese materials, including wax, dyestuffs, cedar, and tallow, and the government would turn its attention to a variety of East Asian plants in the 1850s. But special energy was exerted in the cultivation of tea, which was less labor intensive than rice yet prized as a staple and a stimulant like coffee and tobacco. Agriculturalists in the United States sought to turn useful plants to cash crops for export.

Americans followed British reports with interest, particularly those of Robert Fortune, who documented his travels with some relish.[8] Fortune, a very tall Scottish man, disguised himself as a Chinese merchant, shaving his head and wearing customary dress. He described falling prey to pirates, boar traps, and clumsy barbers, much to the delight of an Anglo-American reading public.

Getting seeds was not easy, which Fortune was quick to observe. Chinese ports restricted the mobility of foreigners. Fortune noted the "jealousy of the Chinese government" in preventing foreigners from visiting any of the districts where tea was cultivated. Local merchants resisted doing business with Fortune. In searching for flower gardens on behalf of the London Horticultural Society, he found the people of Ningbo "unwilling to give . . . the slightest information." They directed him instead to flower shops, and insisted that he had no purpose in a nursery. Fortune found their motives "difficult to define," concluding that it was perhaps "jealousy and fear." On several occasions he described crooked merchants who sold him bad seeds, overcharged him, or lied about the origin of the plants he wanted to buy. There were also reports of seeds being sabotaged, though Fortune tended to disbelieve them. Nonetheless, he avoided trade with an old gardener near Canton who was reputed to boil his seeds so that some "enterprising propagator in England or America" might spoil his business.[9]

Scrappy merchants may not have been averse to making a quick profit from Fortune's ignorance, but resistance to collectors ran deeper than simple hostility or individual profiteering. Merchants were reluctant to give Fortune too much control over the conditions of trade or the botanic resources of their territories. This reflected the kind of savvy that led Hong Kong merchants to boycott trade with the British settlement rather than submit to the British governor's new registration policy.[10] Chinese merchants had no reason to look kindly on a British trader in the wake of an Anglo-Chinese war to secure the future of British opium trafficking.

Regardless of motive or animus, the Chinese tried to protect tea cultivation as a trade secret and succeeded until the British stepped up their efforts to break their monopoly. Transplantation wasn't just about seeds; it was also about technical knowledge. Even as westerners tried to import tea, they lacked information about the plant and its processing.

Cultivating tea required substantial knowledge about seasons, soil, setting, and picking. Processing required still more elaborate production. In China, skilled and semiskilled wageworkers often managed production for merchants who rented land. Peasant producers, too, sometimes hired specialist artisans to help manage production.[11]

Building up European knowledge about cultivation was a slow, flawed process. It relied on previously published accounts of questionable veracity, limited translations of Chinese treatises on tea cultivation, and rare firsthand accounts of tea growing and processing those Fortune provided. His first reports of Chinese tea culture obsessively documented the steps of processing, including repeated firings, siftings, and rolling. Fortune emphasized the skill of Chinese workers.

Even with such detailed reports, British tea prospectors did not assume they could replicate tea culture without oversight. Under direction from Kew's Joseph Banks, the East India Company charged Fortune with enlisting skilled Chinese laborers to instruct planters in Assam. Fortune brought Chinese experts with his plants, where they initiated production on a plantation basis.

The residence of Chinese specialist workers was brief. British overseers reportedly found their émigré workers too particular and proud for plantation labor, ultimately dismissing them in favor of indigenous workers from whom they had appropriated land and resources. Tea producers prioritizing cheap labor first recruited low-wage workers from Assam before turning to coolie workers imported from India. They regarded Assamese workers, including Kachari "plains tribes" cultivators and the Singpho, Khamti, and Naga peasants of Upper Assam, as too truant and stubborn, reading as indolence their resistance to being forced into a cash economy based on opium production for export. Workers conscripted from India, in contrast, lacked social networks and resources to resist disciplinary control once transported to Assam plantations. Low wages, poor quality of life, and harsh supervision were the norm. As Assamese peasants and gentry aimed to separate themselves from the most marginal of workers, coolie became a racial designation in addition to an economic one.[12] The plantation system thus exploited racism as a means of labor control in ways that recalled the enslavement of Africans in the Americas.

British planters and overseers of Assam plantations regularly made analogies between Assam coolies and American slavery, though rarely as a point of self-critique. Planter Alick Carnegie wrote home of the "awful work" they

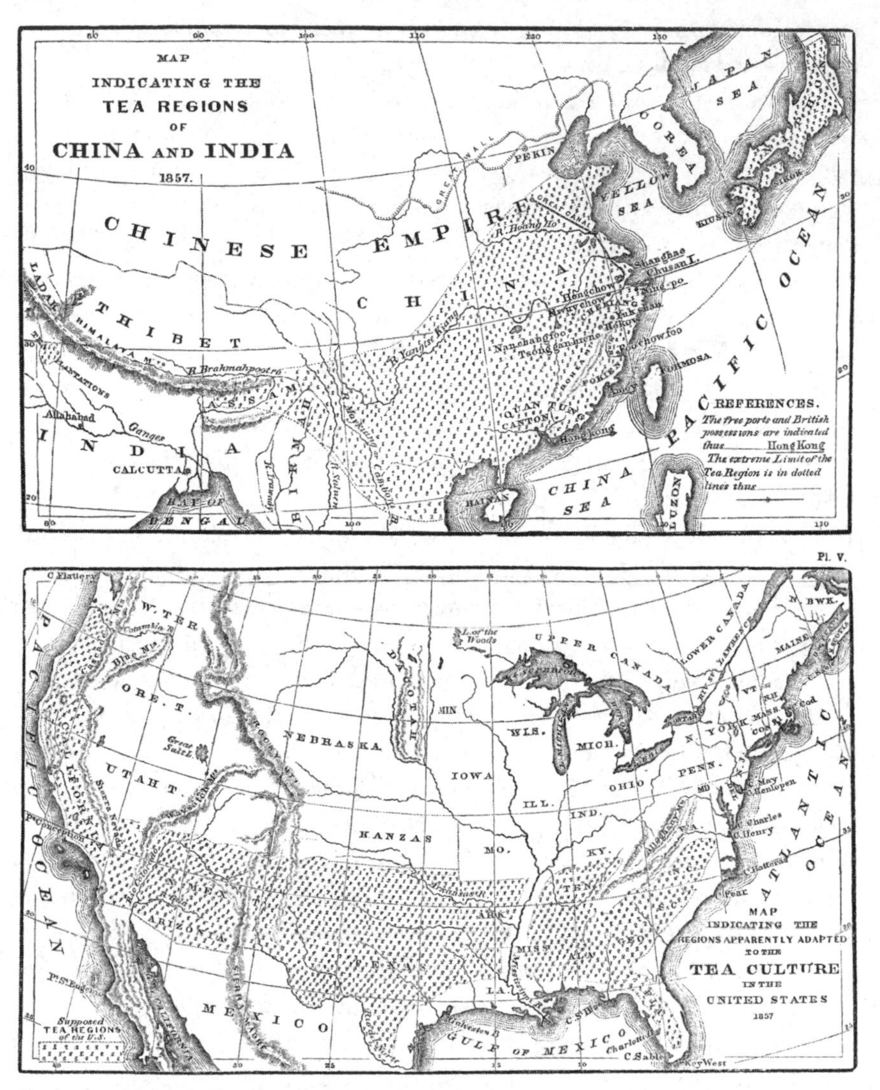

Figure 3.2. "Maps Indicating the Tea Regions of China and India and the Regions Apparently Adapted to Tea Culture in the United States, 1857," published in the Agricultural Report of the Commissioner of Patents for the Year 1857.

had "driving the coolies . . . up and down the line," explaining that they had "to shove them on exactly as nigger drivers in America." One traveler wrote of "notices posted at river ferries and railway stations describing runaway coolies and offering rewards for their apprehension that reminded one of *Uncle Tom's Cabin*."[13] In these depictions, coolie workers were unskilled and hostile bodies rather than knowledgeable workers.

By comparison, Fortune's ethnographies of Chinese cultivation praised the skill of workers. His observations mingled conventions of travel literature, natural history, and horticultural instruction. He gained knowledge about tea culture by observing, recording, and enlisting the bodies and expertise of native cultivators, producing a kind of agricultural knowledge that was highly subjective. Ultimately, self-appointed experts distinguished themselves from one another by claiming deeper and more authentic knowledge of the botany, climatology, and ethnography of native production. Later critics would charge that Fortune was merely a tourist, lacking any sustained experience in cultivation.[14] His critics were right in suggesting the limits of his insights into unfamiliar cultures and environments. His assessments of character were particularly weak and generalized. While preferring farmers to city folk, he nevertheless mocked their fear and fascination with him and prided himself on overcoming their skittishness.

Yet Fortune's reports on cultivation are more technical and more humanistic than his general observations on Chinese life. In the process of observation, he also came to regard family farming with admiration, offering that "there are few sights more pleasing than a Chinese family in the interior engaged in gathering the tea leaves, or, indeed, in any of their other agricultural pursuits." He attributed their contentment to farming for subsistence. "Labour with them is a pleasure, for its fruits are all eaten by themselves, and the rod of the oppressor is unknown. . . . I really believe there is no country in the world where the agricultural population are better off than they are in the north of China," he concluded.[15] If his assessment of family labor and agrarian life was overly romantic and patriarchal, it was nevertheless noteworthy in light of Britain's preference for plantation systems and factory production of commodities.

Tea was well suited to plantation cultivation. Like rubber and other tropical crops, tea trees survived many generations and produced year round. In the newly annexed territory of Assam, British planters took advantage of colonial enclosures of large amounts of uncultivated public land. They standardized plantings, mechanized rolling and firing, and ultimately relied on cheap labor. Produced in this way, tea became a low-cost, standardized commodity for mass consumption. Augmented with sugar produced on West Indian plantations, black tea became the beverage of all British and cheap fuel for the laboring poor.[16]

If Fortune's praise of family farming allowed for a critique of coolie labor in Assam, or legal slavery in the United States, it was one that would not be taken up. Western critics charged that no nation could compete with the cheap labor of the Chinese. British authorities circumvented this concern by instituting a plantation system in Assam. In the United States, boosters understood that the cheap production of tea for world markets required either its incorporation into the slave system or a mechanical solution.

But getting seeds and growing them was the first problem. The United States lagged behind Britain with respect to both botany and piracy. There were small-scale efforts to introduce tea plants into the United States before the 1840s. In 1800, the French botanist André Michaux planted tea near the Ashley River in South Carolina, about fifteen miles from Charleston. In 1817, the *Niles' Register* reported that tea plants were thriving in Virginia. Six years later the same publication reported that "genuine Hyson tea" had been cultivated in North Carolina from a seed found among tea leaves, a dubious proposition at best. A hopeful article in the *Southern Agriculturalist* in 1828 claimed Michaux's plants had been raised for the last fifteen years in Monsieur Noisette's Nursery.[17]

In 1846, one especially ambitious planter, Newbold Puckett of North Carolina, attempted to secure a patent "for the discovery of a mode of cultivating the tea plants."[18] While nineteenth-century patent law commonly excluded rights to products of nature, technical manipulation in the process of cultivation might have been deemed patentable. But Puckett's unsuccessful petition seemed to recall an earlier English or colonial patent regime, which rewarded socially useful industry with monopoly regardless of novelty.[19] Puckett's memorial was ordered printed and referred to the Committee on Patents and the Patent Office, where it seems to have met an end. (No patent appears to have been granted, and the Senate Committee records for the 29th Congress have been lost.)

With the exception of these fledgling attempts, no systematic effort was made to secure seed until Junius Smith, a Connecticut man living in London, turned his attention to tea. The way Smith told it, his daughter was married to a chaplain appointed to the British Army in India, and in the course of her travels to escape the sweltering heat of the Meerut plains, visited the newly formed tea plantations in the Himalayan Mountains. She reported her observations to her father, who received them as a calling, thus unwittingly becoming his tea connection in India. Smith devoted himself to the project entirely: "I did not seek the employment. It sought me," he avowed.[20]

Yet Smith's rhetoric belied the extent to which his scheme was inspired by the popularity of Fortune's endeavors on behalf of the East India Company, and his connections and expertise were inferior to Fortune's in every respect. Smith relied heavily on published resources to build up his knowledge of tea cultivation, a practice that inspired later commentators to charge that he had insufficient knowledge to succeed in growing tea.[21] In 1848, Smith visited the British Library, the London Horticultural Society, and the offices of the East India and Assam Tea Companies, reporting to his nephew that he had "got into the intestines" of the latter establishment and was full of curious intelligence for a pamphlet.[22]

He was as canny about protecting his potential source of plants as his business information, reporting to his nephew regarding seed from China and India: "I do not want the subject to travel beyond our family circle, for I do not want to hunt the squirrel for New York rascals to shoot," he wrote. Yet while Smith played the canny inventor, his trade secrets became common knowledge, and his plants withered in transit.[23]

Smith came up short in land and capital, securing only one small tract of land in South Carolina rather than the fourteen sites he had planned. And though he persisted with monthly plantings to ascertain ideal times for germination, he heard his neighbors say his project was "a total failure." He groused that South Carolinians knew "nothing but to plant corn and cotton" and protested that the public was "naturally impatient of delay."[24]

Others endeavored to succeed where Smith had failed in getting reliable seeds and cuttings directly from China. The same 1850 annual report that had lauded Smith's gradual progress and noted experiments in Brazil also included a report on tea in Assam forwarded by J. Abbot Lawrence, US minister in London. The author was Francis Bonynge, a former agent of the East India Company who boasted fourteen years living in China and north India (as well as being forcibly driven from his plantation by Singpho residents from whose land it was carved). Determined to leverage his expertise for financial gain in the United States, Bonynge made overtures to American agriculturalists, prepared a prospectus on the cultivation of tea and other Asian plants in the United States, and took to the road in 1851, touring the Southeast in search of customers. He promised those who enrolled twelve tea plants and a series of other tropical plants, including mango, date, lychee, and coffee. Of these, tea attracted the most attention. Fifty dollars would secure the plants; one hundred would guarantee any other plants from India the client might desire, along with an expert in tea cultivation.[25] Here again, in the absence of knowledge about cultivation, workers figured as experts.

Bonynge operated as a kind of niche seed firm with better suppliers than Smith had. He offered seeds direct to farmers on a subscription model. He recognized the difficulty of procuring seeds, and perhaps he even exaggerated it for the benefit of his potential customers. He assured American readers that they had parity with the British East India Company as to the difficulty of acquiring plants from China. "To get good seed," Bonynge claimed, was "not to be accomplished. The East Indian British Government (and no party had the same opportunities) could not succeed." The Chinese, he claimed, gave foreigners seed only from the stiff hard soil of Amoy and Quang-Tong, which produced tea "of such inferior quality that natives do not manufacture it beyond the simple drying of leaves." This inferior seed was "often mixed with good teas and sold to England and America." The proper tea seeds, in contrast, were found some 1,000 miles from Canton, or 260 miles from Shanghai of difficult overland traveling. Bonynge claimed that all the seeds Fortune conveyed to British India could not produce tea except by stripping off all the leaves and thus denuding the plant of many seasons' growth. Bonynge promised to secure the best seeds for his subscribers.[26]

Bonynge also boasted of his knowledge as a skilled cultivator. Although he cited Fortune's observations on China repeatedly, he derided his overall knowledge of tea cultivation. He noted that Fortune had been placed in charge of a failed plantation in Kamoun in the northwest of east India, but that he had "no further knowledge of tea matters."[27] Bonynge, in contrast, had studied tea cultivation in both China and Assam, giving him the alleged advantage of viewing variable conditions of climate and labor. Bonynge charged Smith with relying mainly on previously published sources gleaned from the London Horticultural Society and the British Library. Future critics would echo his suggestion that printed accounts were insufficient basis for successful transplantation.

Bonynge's prospectus, self-published in 1852 as a treatise titled *The Future Wealth of America*, included testimonials and letters of recognition from a number of prominent politicians and agriculturalists. Daniel Lee, editor of the *Southern Cultivator* and an agent of the Agricultural Division of the Patent Office, publicized Bonynge's efforts and expertise in the pages of his magazine. While the British had succeeded in India, he noted, previous importations of seed by the Patent Office had not vegetated. Lee cited Fortune's success in procuring plants for Assam and suggested that Chinese migration to California was an opportunity to capitalize on their experience in tea culture and the speed of the Shanghai to San Francisco trade route.[28] Lee was probably partly responsible for Bonynge's inclusion in the 1850 Patent Office report alongside Smith.

But in spite of his testimonials and support in high places, Bonynge's subscription model of plant introduction fizzled. In July 1852, Bonynge presented through Senator Thomas Jefferson Rusk of Texas a memorial requesting an appropriation to enable him to introduce and cultivate tea, indigo, coffee, and other tropical plants and fruits into the United States.[29] The memorial seems to have died in committee. In 1854, *Scientific American* reported "a gentleman passing through Dunkirk en route for Cincinnati with some 12 Chinese culturalists, for the purpose of testing out the practicability of growing tea in the vicinity." This may have been Bonynge. The author noted that nothing had been heard on the subject since the death of Junius Smith of South Carolina.[30]

Although the Patent Office's Agricultural Division, through Daniel Lee, had supported Bonynge's enterprise, ultimately both he and Smith had to compete with its superior resources. Initially, the Patent Office's propagating garden had only a modest collection of tea. William R. Smith's account of the botanic garden, included in Alfred Hunter's catalog of the museum in the Patent Office, noted isolated specimens of tea, along with camphor trees from Japan, bay laurels from Italy, and lemons.[31] Smith singled out the tea plant for several observations. He noted that specimens of the Chinese species *thea bohea* produced both black and green teas, that another called "Thea Assamica" furnished Assam tea, and that the exact number of tea species was unknown.[32] Smith claimed to rely on the best botanical authorities for his catalog, and indeed, at the time of publication, information about the tea plant was far from commonplace.

Then beginning in 1849, Samuel Wells Williams of the American Board of Commissioners of Foreign Missionaries devoted his energy to the Patent Office's efforts to secure tea plants. He enlisted his connections in China to collect many kinds of plants and transport them to the United States via navy and merchant ships associated with the US legation at Macao.[33] Williams argued that only government aid, exploration, and education could ensure such an ambitious project's success. To support the effort, Williams enlisted his missionary connections in China. They in turn relied on the expertise of Chinese physicians.[34] With seeds, Williams and his connections also conveyed detailed observations about Chinese agricultural practices.[35] The commissioner of patents in turn distributed the information to select farmers in Arkansas, Florida, and South Carolina, including Junius Smith, who was by then a regular correspondent of the Patent Office.[36]

But Smith was disconsolate. Gratis distribution of seeds and agricultural knowledge threatened to render his contribution to American agriculture irrelevant. In the spring of 1850, he protested to Edmund Burke, the current commissioner of patents. Having just answered a congressman's request for tea nuts, he explained, the recipient sent another note saying he had already been supplied. "I know he is supplied gratuitously by the Patent Office," Smith complained. "I think upon reflection the government will perceive that the gratuitous distribution of tea nuts is a great injustice to me. I have labored more than three years to introduce tea into the United States. I cannot obtain seeds except by paying, and looking to the public for remuneration; but no one will pay me as long as they can obtain supplies from the Patent Office for nothing. I am a ruined man if the government continues."[37]

Smith perceived the government's involvement as a violation of his rights as an innovator. Having taken great pains to secure a rare agricultural product and build up knowledge about it, Smith believed he deserved the right to sell it with some measure of exclusivity. Although Newbold Puckett's petition for a patent might have riled him had he known of it, in fact he was angling for similar privileges.

Burke did his best to soothe Smith's anxiety by soliciting and publishing select of his remarks on tea culture and lauding him as an "introducer" and "benefactor" with "enterprise and talents deserving of remembrance as established in this publication."[38] Publication in the annual reports of the Patent Office on Agriculture was a source of prestige for agriculturalists. The reports achieved wide public distribution through the offices of US congressmen.

Meanwhile, the Patent Office continued to import seeds through military and consular channels. Dr. Macgowan, Williams's connection in China and an established correspondent of the Patent Office, continued to forward seeds and seedlings, though he complained that the meager sum provided by the Patent Office was insufficient to procure much of value.[39] Meanwhile, Williams would serve as an interpreter on the Perry Expedition to Japan in 1853 and secretary interpreter for the 1857 Treaty of Tianjin extending American rights to trade in China. In practice, free commerce required substantial government intervention, and the agents of trade also pursued nature collection oriented toward strategies of import substitution.

By the conventions of agricultural innovation, Smith deserved praise and publicity, but not rights or remuneration. Smith dutifully conveyed a full edited manuscript of his study of tea cultivation along with engravings of tea plants gathered from the British Library, but a year later, when his engravings had been neither published nor returned, he wrote tensely that he would withhold further correspondence until his originals came back. While the Patent Office's neglect was probably a matter of bureaucratic oversight rather than intentional disregard, it did nothing to encourage Smith's pretensions to success, and his bargaining power in withholding future communication was dubious at best.

It all came to a sad end. In the winter of 1851, Smith's labors were cut short by a violent robbery. The head wounds he sustained surely contributed to the deranged state in which his nephew found him some months later. Conveyed to New York, he died in an asylum in Queens the following year. Months later, the whimsical journalist from *Harper's* found his unclaimed tea nuts languishing in the hall of the Dead Letter Office in Washington.

Smith was an early opponent of the free federal distribution of seeds. By the later 1850s, northeastern seed dealers and agricultural journals distributing free seeds to subscribers complained that federal plant introduction programs were gratuitous and wasteful, in fact responding to encroachments on their own business. By the twentieth century, the seed lobby would advocate patent rights for improved breeds on the grounds that modification constituted invention, culminating in the Plant Patent Act of 1930. Smith's objections presaged these arguments.

Jacksonian leaders accepted the admissibility of monopolies only by rationalizing their protection of truly new knowledge. Smith's protest, anticipating those to come, treated seeds and plants as private property, if no longer subject to patent protection for public utility, at least subject to the rules of free commerce—albeit free commerce that relied on the military enforcement of a British presence in China. At the source of the seeds and in their new homes, trade was restricted by inclinations to protect rather than share natural resources.

That Junius Smith's primary competitor was the federal government expresses yet more instability in the political economy of the Jacksonian period, and specifically, entrepreneurial ambivalence about the role of the Patent Office in a growing economy. The Patent Office's intervention in matters of commerce might be perceived as stimulus or interference, depending on the interests or ideology of the party concerned.

Nor was the Patent Office clear on the question of which systems of production it supported for tea culture, rendering its ideological position still more uncertain. Some suspected that Smith's antislavery sentiments contributed to the attack on his person. Smith believed the key to cost-effective production lay in transportation. Having spent the last ten years advocating transatlantic navigation by steamship, he believed that with internal improvements, steamships would dramatically reduce the cost of bringing tea to market.[40] Superior infrastructure would make tea culture into an industry proper, and Yankee ingenuity would perfect the culture of tea.

Others emphasized the power of labor-saving machinery to simplify the sorting and rolling of tea. Spencer Bonsall, a Philadelphian formerly employed on the Assam tea plantations, claimed steam-heated metal plates and circular wire screens moved by steam power could dispense with the labor-intensive hot-hearth and sifting processes used in China.[41] "Necessity is the mother of invention," avowed one agricultural journalist, "and a relation of that family, an acute son of New England, has already set his mind upon a tea-curling machine which promises to do for the American crop with a few thousand fingers of steel, the work which occupies the digits of a million inhabitants of the Flowery land."[42]

Meanwhile, southern boosters emphasized the continued viability of slave labor as a solution to the labor-intensive requirements of tea cultivation. Francis Bonynge courted southern planters with comforting assurances about the institution of slavery. An expert in plantation governance, he assured planters that though legal slavery might be destined for obsolescence, it was no great matter: "when the British East India Company say they abolished slavery in the East," he explained, "they did nothing more than pass an act against a term that had no meaning. Among all the higher classes in India today, there are slaves—who are so, willingly; who might be more appropriately called hangers-on, because they cannot do better."[43]

In the absence of legal slavery, de facto slavery would do. The southern agricultural press, led by Daniel Lee, failed to appreciate Bonynge's brutal subtlety, promising instead that tea would flourish through the agency of slave labor. "I feel warranted in expressing the opinion," he avowed, "that the time is not far removed when Southern enterprise and field hands will excel the Chinese as much in the simple operation of picking and curing tea leaves, and growing the trees, as they do now in growing, picking, and ginning cotton."[44]

The Patent Office retained its interest in tea cultivation and its reluctance to consider its proper political economy. In 1856, Daniel J. Browne, Commissioner Mason's agricultural chief, prepared a lengthy prospectus on the introduction of tea culture to the United States, stressing again the climatic similarities between East Asia and the southern United States, now vastly expanded to include territories annexed in the Mexican War. Browne's two-part prospectus appeared in the 1856 and 1857 agricultural reports of the Patent Office.[45] He drew on major efforts to date, including Jameson's account of superintending

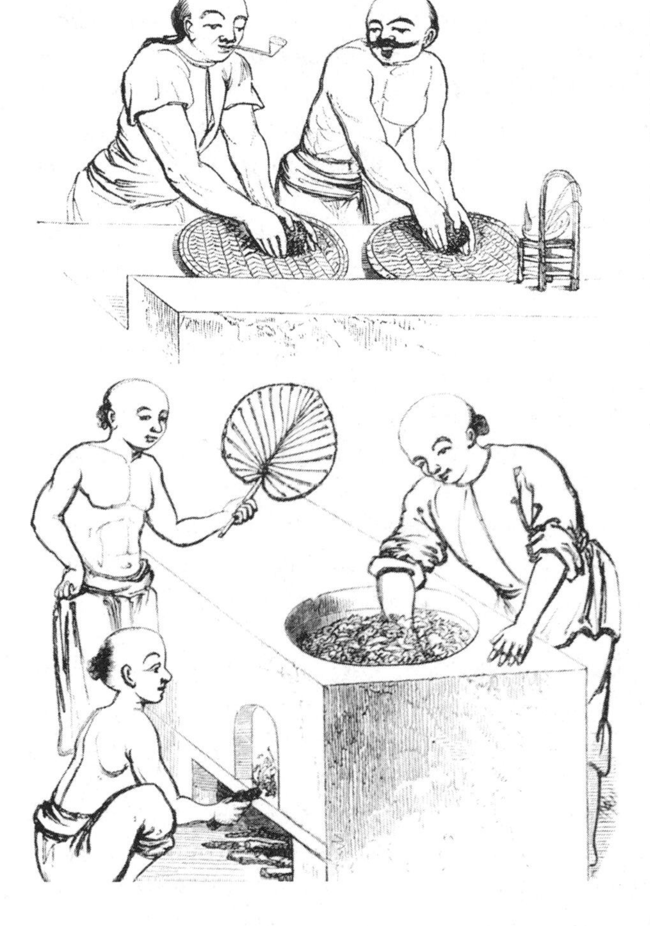

Figure 3.3. ***Harper's*** **reprinted the Patent Office's maps of tea regions from the 1857 annual report and drew heavily on Robert Fortune's travel narratives, describing the processing of tea at length. Here, rollers squeeze juice out of tea leaves on bamboo mats and fan leaves during roasting.**

the botanic gardens in Assam, and Fortune's several published works on the tea countries of China.

Browne answered the labor question with by now customary assertions: internal improvements in transportation would speed tea to market. He repeated the argument of Bonynge and Kew botanists that American labor was more reliable and robust than that of Asiatic bodies. And he relied on Bonsall's speculations regarding labor-saving machinery. American ingenuity would easily produce labor-saving machinery that would do the work of one hundred men.[46]

His arguments, along with meteorological analysis performed in collaboration with the Smithsonian Institution, set the stage for Commissioner Holt to hire Robert Fortune, by then director of the Chelsea Physick Garden, London's oldest botanical garden and a storehouse of global horticulture. In 1857, the Patent Office approached Fortune through the London seed firm of Charlwood & Cummins. Fortune forwarded his opinion on the probability of securing tea plants. He advised that the best route would be to follow the example of the East India Company and introduce the best varieties of tea from China directly, agreeing to enter into an arrangement with the US government similar to the terms he had with the company. For a salary of five hundred pounds per year plus travel expenses, Fortune agreed to forward to Washington glass-vented cases of living tea plants according to the method he had outlined in *Journey to the Tea Countries of China*.[47]

Fortune established himself at the outset as an expert in cultivation, not just a seed hunter. He explained that the office's proposal to send seeds in cases would not succeed, since the seeds lost their vitality out of ground. He referred them to his opinions on the subject from his most recent publication. His eventual agreement with the Patent Office in November 1857 also charged him with collecting other "plants of China which would be likely to suit the climate and soil of the US and to visit the US upon his return from China, for the purpose of selecting proper sites and of giving instructions as to the future management of the productions" he might introduce.[48]

So he was unpleasantly surprised when, a year and a half of travel and half a dozen shipments of plants later, Holt informed him that his services were no longer required. Fortune repeated a now customary argument regarding the necessity of technical expertise. He wrote to Holt that it would have afforded him much pleasure to have given the office "the benefit of his experience in rearing and transplanting to proper sites the Tea and other useful productions," adding that it would be "a source of deep regret if the experiment should fail from want of experience which can only be acquired in the country to which these plants are indigenous." He demanded severance, and having received it congratulated Holt, noting with more self-aggrandizement than praise that more had "been done in one year for the US than was done in three for the government of India." On the commissioner's request, Fortune provided written answers to questions regarding the proper cultivation of tea.[49]

Holt evidently concluded that having gotten the plants in living condition and in light of the extensive publications on the subject, Fortune's guidance was not required. Times had changed.

On the eve of the Civil War, southern cultivators enthusiastically reported that genuine Chinese tea plants of seven years' growth were flourishing in the nursery operated by the Patent Office constructed from First Street to Third Street between Pennsylvania and Maryland Avenues. "The new propagating Garden is operating to a charm," one correspondent reported to the *Southern Agriculturalist* in August 1859. "There are already growing in the green houses over 50,000 Tea plants, and more seeds and plants are on their way from China."[50] Abutting the collection of tea plants were grapes, figs, and pomegranates from Egypt, along with wax trees, camphor trees, loquats, lychees, and other specimens from China and Japan. The greenhouses and the adjoining garden served as a staging area for specimens prior to distribution.[51] By the spring of 1860, tea seedlings had been distributed to farmers in southern states.[52]

But tea did not flourish for local use or the domestic market, nor did it become an export to rival cotton or wheat. The Civil War interrupted the experiments. The Patent Office's agricultural agents turned their attention to gathering specimens of cotton from behind Confederate lines and shipping them west to the new states of the Union, rewarded for their loyalty with seeds taken from rebels. This system of rewards signaled a lack of commitment to previous and subsequent disquisitions about diversifying southern agriculture. Meanwhile, India took its place on the world stage as a center of tea production, surpassing Chinese exports by the 1880s and making tea a beverage for all Britain.

Experimentation continued after the Civil War, but sporadically. The newly founded US Department of Agriculture continued to promote tea, remade briefly as a crop to reconstruct southern agriculture around family farming during the Civil War, then again as a source of labor for women and children. Fresh supplies of seed from China and Brazil signaled the office's continued commitment to tea cultivation, but attempts were made in fits and starts and according to the whims of successive commissioners of agriculture.[53]

The most successful endeavor was Charles Shepard's Pinehurst Plantation in South Carolina. Shepard initiated plantings in 1888 with seeds and cuttings gathered from the failed plantations of John Jackson, who had tried to cultivate tea in Georgia and Summerville, South Carolina, in 1880 and 1881.[54] In 1902, the *New York Times* proudly reported that a delegation from Japan and University of Tokyo scientists took great interest in the operations of the Pinehurst Tea Farm.[55] That year Shepard also submitted a celebratory entry on American tea for Liberty Hyde Bailey's *Cyclopedia of American Horticulture*.[56] But Shepard's death, like Smith's before him, left his plants unattended for decades. Lipton acquired the plantation in 1960, opening a research station on Wadmalaw

Island, but sold it to a private proprietor in 1987, from which time it has been operated as the Charleston Tea Plantation: producer of American Classic Tea, supplier for the White House, and recently stocked at Wal-Mart.[57]

While China lost its monopoly on tea production in the late nineteenth century, Fujian nevertheless continued competitive production for world markets of high quality teas, retaining a model of decentralized, smallholder production. These farmers continued to produce other crops and food for subsistence. Skilled wage and family labor suited the diversity of green tea products for Chinese markets.[58]

In contrast, English plantations in Assam and Ceylon prioritized strict labor control and plantation production of a uniform commodity for mass-market consumption. Rather than a diverse group of commodities serving many cultural functions, tea spiked with sugar became a drug food for all Britain.[59]

Expert knowledge was thus not simply a factor of production but also of consumption. Taste preferences were conditioned by prior exposure and use. Meanwhile, skilled labor combined botanic and agrarian knowledge with elaborate styles of production. In British and American transplantation efforts, there was a vacuum of connoisseurship at all levels. British botanists looked askance at Assamese tea plants on the grounds that they were too wild, then instituted colonial production on the basis of these racist fictions, alternately hiring and firing Chinese experts, recruiting and deriding indigenous workers, and conscripting coolie laborers on the assumption that they neither possessed nor required any knowledge of the material they were to produce. Americans, modeling their efforts after the British in Assam, lacked the knowledge to produce anything other than a mass-market commodity, thus ignoring the ways it reduced many centuries of technical knowledge and practice to a single trade article. Just as the imperative of cash crop production deskilled workers, it also deskilled consumers, who knew only black tea, not the numerous preparations of ritual and diplomatic importance in China.

After Indian independence, migrant workers in Assam remained a permanent laboring class, a legacy of their displacement and racialization.[60] Meanwhile, Assamese "tea tribes" have recently battled over who deserves the title of the first indigenous tea planters, with the Singpho claiming priority over Maniran Dewan, for whom the local planning commission named its regional tea center.[61] Both narratives of heritage ignore that tea forests had flourished in Assam long before they were rationalized as British-run plantations, and that the latter succeeded on the basis of expropriating local resources and exploiting local people. Tribal identities associated with tea culture thus mask the continued exploitation of labor capitalizing on conflicts between Assam and coolie populations.[62]

This scramble for indigenous claims to rights granted by British plantation overseers in pursuit of a global tea industry reflects the contemporary organization of agriculture around proprietary claims rather than colonial rule. Yet these too mask systems of labor control and economic exploitation. Assam's "tea tribes" claim priority in the discovery of indigenous tea plants, as well as the cultivation of still other plants gotten from China in a plantation system of labor. These support local claims to ownership of a natural product that is also a global brand. Assam and Darjeeling tea makers are also working to acquire a species of intellectual property protection for their products known as a geographic indication, designating a product that could not be made in the same way anywhere else because of biological, environmental, and cultural characteristics. The geographic indication protects much like a trademark to a brand; for example, Champagne, or Darjeeling Tea. But not South Carolina Tea.

The Patent Office, adopting a logic of cash crop production that prioritized the deskilling of labor for the scaling up of production, assumed Fortune's ethnographic knowledge of tea cultivation was unnecessary after his successful conveyance of live seedlings. In doing so the office adopted a blinkered view of agricultural practice, operating on the assumption that seeds were effective substitutes for agronomic knowledge, easily turned to many forms of production.

In fact, it was the unwillingness or inability of agriculturalists to tackle problems of labor and political economy that rendered their experiments in tea cultivation ineffective. The Patent Office advocated neither plantation slavery nor smallholder production such as that which persisted in China. Its leadership remained curiously silent on questions of production, succumbing to the palatable fiction that mechanization could dissolve conflicts between ideologues of slavery and free labor. In the absence of any stronger conviction, farmers planted more cotton.

With another combination of land, seeds, culture, and labor, Carolina tea might have been a triumph of American nationalism, protected by a global brand or a species of intellectual property rights. But instead, it was a patsy to King Cotton, and Junius Smith's unclaimed seeds moldered in the Dead Letter Office, and the remnants of his tea plantation became an accessory in the president's kitchen: a small victory of branding, but not enough to make a classic. In the United States, improvers ultimately failed to envision alternative models of production based on smallholder cultivation, skilled labor, and diversified products, cleaving instead to a plantation system forged in the production of cotton. Ultimately, the myopic focus of improvers on the acquisition and distribution of seed reflected an unwillingness to connect agricultural improvement to questions of political economy. Improvers paid little attention to the colonial origins of the cultivation they pursued, and in doing so allowed its legacies to persist in twentieth- and twenty-first-century agricultural science.

Field Notes

"LOCAL KNOWLEDGE"

WHAT THE PASTORALIST KNEW

Contemporary development and preservation projects may proceed more cautiously than nineteenth-century ones, mindful of imperial legacies and histories of error. Yet in their attempts to cultivate awareness of local ecosystems and environmental knowledge, they face persistent challenges. What is a useful plant? Collectors may have different interests than the farmers from whom they collect. To whose services are plant genetic resources amassed and deployed? Moreover, environmental knowledge is no more stable than the environment itself. Rather, it moves and changes along with the people who steward it, who may face environmental, political, or economic crisis.

Zorats Karer and the adjacent wheat fields in the Armenian countryside in which the team I accompanied traveled provided indications of long-standing settlement. Although we saw no one, the ruins and the plants told us where to collect. Yet we knew, as did collectors of the preceding centuries, that seeking varied plants requires the knowledge of varied people. Knowledge about where to collect seeds comes largely from within target regions, not outside of them. Because plants that can withstand salty soil may fare well in a warming climate, for example, investigators of climate-hardy crops seek samples from saline areas. While researchers come armed with a variety of geographic and climatic data, local people may identify target areas with the most efficiency and precision.

Bound for salty soil, our team traveled to marshy lands in the Aras River valley, twenty miles west of Yerevan, Armenia, and ten miles from the Turkish border. The land is agriculturally rich but also earthquake-prone, situated on a major fault line eerily flanked by Mount Ararat on one side and the Medzamor Nuclear Power Plant on the other. For its cooling operations, the latter draws water from the Aras River, which also marks the Turkish-Armenian border. Control of water resources is a major point of contention for pastoralists and farmers near the border. Cattle grazed there, innocent of their surroundings.

It was the cows that obliged us to stop the van and collect, uninterested as they were in clearing the road for traffic. As we collected, we met the man

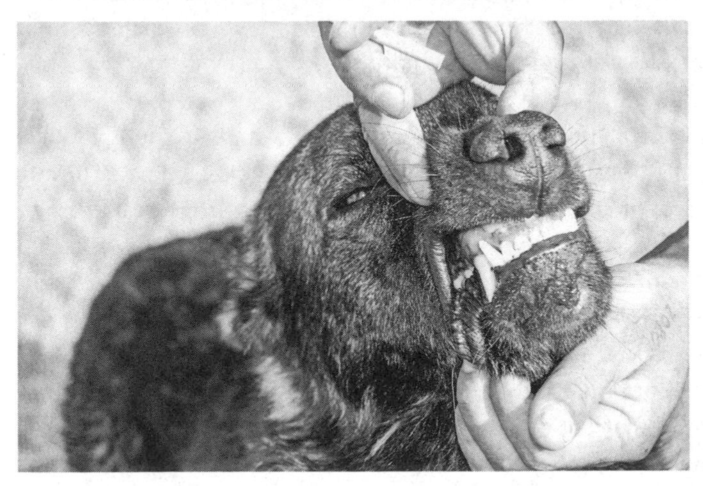

Figure FN2.1. ***A–C.*** **Photos by Courtney Fullilove.**

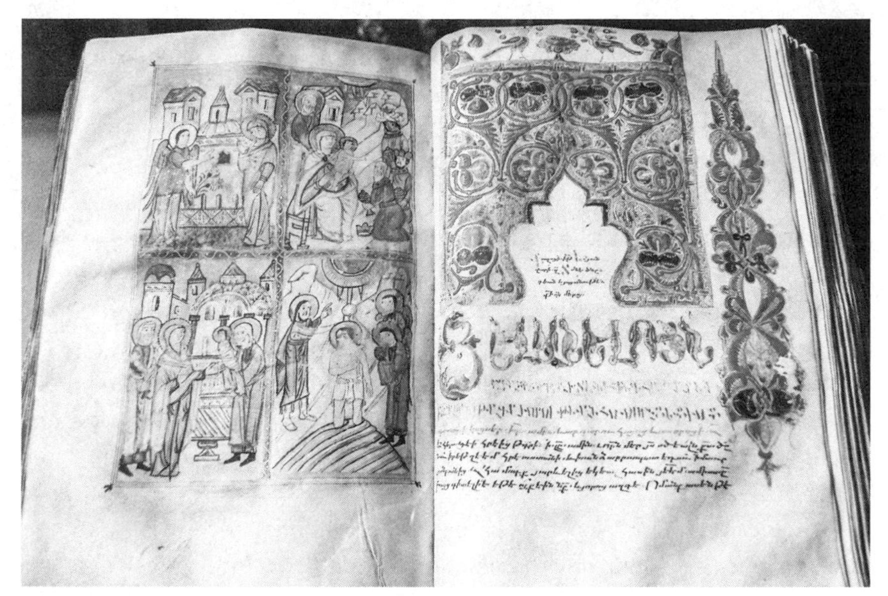

Figure FN2.2. Manuscript on display in the Holy See of Echmiadzin, Armenia.

responsible for the cattle, meandering in the field nearby with his dog, Tonic. An infectiously curious person with an easygoing demeanor, the cattleman readily engaged us in conversation about our work, wondering who we were, why we were there, and what we hoped to accomplish by gathering local flora.

Amiable as he was, he wore his difference clearly. He told us he trained the dog to understand only Yazidi, not Armenian, in an effort to prevent trickery and theft. It was a defensive measure born of experience. He had taught Tonic's mother, Gin, to sniff out mushrooms, extending her value well beyond her cattle wrangling abilities, but jealous neighbors had lured and ultimately poisoned her, he said, leaving Tonic an orphan, too young to have acquired the skills of his mother.

Hearing of our interest in local flora, the herder took us on an impromptu tour of the fields, pointing with the switch he carried toward a nondescript clearing beneath Mount Ararat, which loomed on the horizon across the Turkish border. In the fall, he told us, the field would be red, stained with the blood of worms that had fed on the grass there. He first noticed the place because Tonic would come back from the field with his paws dyed crimson. No one paid attention to the spot anymore, but it used to be worth something. The blood was of a type similar to that harvested for Ararat cochineal dye, also sometimes referred to as Armenian Red, which illuminated manuscripts lavishly displayed at the nearby Holy See of Echmiadzin.

What the herder found important was that most people thought the worms fed on only one kind of plant, but the one in this region was a slightly different variety about which nobody knew. If we were interested in valuable plants, he suggested we collect it and cultivate it.

But we were interested in wheat, not dyestuffs, and people had long since ceased to produce the dye in any case. Now the dye's only purpose was to decorate Tonic's paws as he traversed the fields. For that matter, few remember the location or function of the worms' blood, so it was left to the pastoralist to think about this lost knowledge as he wandered between Ararat and the nuclear plant.

As our team wound north through the mountains to the Georgian border, I thought about the man and his dog walking through the countryside, and about what sorts of knowledge about mushrooms and dyes they accumulated along the way.

In a vocabulary increasingly shared by international development theory, a history of science, and environmental studies, this is "local knowledge," experiential knowledge inseparable from practical skills, particular to local cultures and environments, and apparently common to people in a given community.[1] The category of the local is not restricted to indigenous people, nor is this category of knowledge properly defined by the traditional, isolated, or unchanging. Rather, it embraces perceptions, practices, and ideas possessed by any social group, acknowledging the possibility of exchange within or among communities. The openness of this formulation challenges the notion that rural, agro-ecological, or craft knowledges are endemic to native people bearing them changelessly across time.

But what about people on the move: willing and unwilling migrants, displaced people, pastoralists, and nomads? Anthropologists have identified these as people resistant to governance, perhaps sharing tactics for evasion and autonomy from the state with peasants worldwide.[2] In the pastiche of tourist literature, these traits form part of a romance of mountain people, spiritually independent and defiant of rule, yet ultimately facing destruction or assimilation: the endangerment of the pastoralist is part and parcel of his appeal. Preservation derives both its drama and its moral force from this imaginary.

Such fairy tales sidestep questions that are at once more ambiguous and more pointed. Who or what endangers? Development? Conflict? Commerce? Government? Who or what is endangered? Biodiversity? Culture? Ecology? Humanity?

It is tempting to regard the pastoralist as a source of tradition or indigenous knowledge, but in fact the very opposite is true. Numbering perhaps half a million, the Yazidi are centered in Iraqi Kurdistan, with diaspora populations in Syria, Turkey, Armenia, and Germany. In August 2014, many became aware of the plight of the Yazidi community in northern Iraq, trapped on Mount Sinjar after fleeing attack by the Islamic State (ISIS), a jihadist group with territorial aims in Syria and Iraq. An ethnically Kurdish minority whose religion blends elements of Islam, Christianity, and Zoroastrianism, Yazidi people have struggled even prior to the recent attacks, displaced in turn by exclusion, alienation, and outright violence. This man didn't have knowledge because he was indigenous or traditional. He had knowledge because he was a traveler.

The area around the Jrarat and Arazap villages in the Vagharshapat region of Yerevan are indeed home to Ararat cochineal (*Porphyrophora hammelii*), an endemic insect species producing carmine dyes found near the foot of Mount Ararat in Armenia and Turkey. The dye insect grain cochineal (*Porphyrophora tritici*), in contrast, was native to central Anatolia until likely being eradicated by the DDT used to protect the Turkish wheat crop from decimation by the insect in the 1960s. Scarlet textiles from Ararat are present as early as the eighth century BC in inventories of the Assyrian king Sargon II's plunder from his defeat of the Urartean Kingdom. First- and second-century Roman textiles in Palmyra contain carminic acid found in these insects, and Armenian literary sources from the fifth century onward describe their use in silk dyeing and miniature painting. Several contemporary Armenian institutions have attempted to revive the production of the dyes, and these sites are visited in the fall by small eco-tours organized in Yerevan, which also visit nearby cathedrals dating from the fourth through seventh centuries.[3]

There is no reason to doubt that Tonic discovered the dye for himself, wandering the fields, and so perhaps an orphaned dog deserves all credit in the matter rather than any human. On the other hand, Tonic's Yazidi guardian may have learned as much about these insects from chemists and tourists as from his dog's initial collection of the worms' blood.

A clear-sighted assessment of local knowledge should allow for its mobility and alteration in response to environmental and social interactions. More cosmopolitan than local, the trails of people and animals trace knowledge on the move. Rather than capitulating to the romance of endangerment, we need to ask how plants and knowledge about them are transformed by the more quotidian business of daily life.

Our histories, too, should account for ecological change and the effects of knowledge on the move. Histories of ecologies and environmental knowledge may require departures from traditional geographic or political frameworks.

For example, what would a history of grasslands look like? How have migration and settlement altered environmental relations between human and nonhuman actors?

The grazed lands at the foot of Ararat bear some similarities to the steppe sweeping from the Ukraine and North Caucasus to Manchuria, and for that matter, to the Great Plains of the American Midwest. For each is a large area of flat, unforested grassland with a semiarid climate. Moreover, large areas of the southern steppe and the Great Plains, both characterized by humus-rich black soil, were transformed in less than a century from wild grass plains to sites of monocultural grain production.

The Great Plains cover three million square kilometers, stretching from Texas to Saskatchewan and from the Rocky Mountains to the forests of Missouri, Indiana, and Wisconsin. The prairie, consisting of tall grass in the east, mixed grass in the center, and short grass in the west, hosts varied flora, including numerous grasses and wildflowers that flourish in arid environments and full sun. The stiff soil cover of the prairie inhibited cultivation until the mid-nineteenth century, when steel plows and other agricultural implements aided westward migration of Euro-American farmers.[4]

But neither large-scale grain production nor more modest agricultural settlements were the first transformation of the grasslands, which had been converted by the mid-nineteenth century into an energy source for herbivores supplying trade in horses and livestock. In earlier centuries, grazing bison, antelope, and elk prevented extensive tree growth, as did drought. Native American settlers inhibited forestation through fires.[5] Pekka Hämäläinen has examined how Comanche Indians settled in the plains adapted to an environment transformed by European plants, animals, and microbes. As they expanded their use of horses for bison hunting in the eighteenth and early nineteenth centuries, the Comanche increased trade relationships based on exporting horses, hides, and meat. Their turn to horsepower intensified the use of plains grasses for human use. The Comanche experimented with transhumance and frontier raiding to retain their equestrian economy's grip on the plains. In the short term, prairie grasses that had coevolved with North American big game proved resilient to heavy grazing. But by the mid-nineteenth century, market-oriented pastoralism faltered in the face of concerted government commitment to Indian removal and agricultural settlement of the American West.[6]

Many witnessed this transformation, yet few apprehended its full implications. The historian Frederick Jackson Turner regarded the American West as the frontier, a savage wilderness transformed by settlers of European descent into a distinctively American form of government. Reading the figures of the 1890 census and noting the reach of Euro-American settlements from the Eastern Seaboard to the Pacific, he wondered what would be the next stage of American democracy.[7] Theodore Roosevelt too regarded the West as the proving ground for American nationalism and commercial empire, and he admired Turner's thesis. Yet whereas Turner looked to the future with some temerity, Roosevelt considered the "winning of the West" a preparation for the extension of American empire in the world.[8]

Some weeks before Turner took the podium at the American Historical Association conference in 1893 to declare the frontier closed, Wellesley College English professor Katherine Lee Bates boarded a train to teach the summer session at Colorado College. Gazing out the window, she found herself so moved by the wheat fields of the Midwest that she put pen to paper and composed the opening lines of a poem: "O Beautiful for spacious skies / For Amber Waves of Grain / For purple mountain majesties / Above the fruited plain."[9] Set to music by Samuel A. Ward, the opening stanzas became a de facto national anthem, in which both geologic features and agricultural products figured as permanent aspects of the landscape and embodiments of national patrimony. The prosperity she celebrated represented the triumph of manifest destiny and the fruit of an agricultural empire that spread westward across the continent.

Yet the amber waves of grain Katherine Lee Bates made iconographic of the American landscape were not native to it, or indeed more than twenty years old. The wheat plant itself was a transplant from northern Europe, and the several varieties that made the Great Plains a breadbasket to the world were conveyed there only in the 1870s, reputedly by German Mennonites emigrating from southern Russia, in what is now eastern Ukraine and Crimea. For all its romance of national bounty, Bates's verse was tantamount to celebrating American heritage in silver seas of iPhone.

The history to which Bates referred was one of movement rather than stasis, and innovation rather than tradition. Although her poetry would be put to the service of ideological narratives of civilization and progress, it obscured a more complex history of migration and displacement in the production of knowledge, ecology, and political economy. This history of immigrant agricultural knowledge on the nineteenth-century American prairie is the focus of the succeeding chapters.

PART 2 : MIGRATION

WHEAT CULTURE AND IMMIGRANT AGRICULTURAL KNOWLEDGE

(Peabody, Kansas; Molochna, Taurida, 1871–1903)

4 : FOR AMBER WAVES OF GRAIN

The night after Bernard Warkentin first laid eyes on the Kansas prairie, he wrote to childhood friend David Goerz in Crimea of the fields strewn with bison carcasses he had found there. Thirty years later, his traveling companion, Christian Krehbiel, still remembered the half-decayed bodies and "enormous bone heaps of the same animals, piled up in the towns."[1] But the two were also struck by the beauty of the landscape, the carpets of wild flowers and grassy cover reminiscent of the uncultivated steppe Warkentin's Mennonite ancestors found in southern Russia only seventy-five years before. Like the steppe, the prairie was not terra nullius, but rather landscape marked by successive removals of people and animals. In fact, the prairie trails the new human visitors traversed had been carved by the very animals they slaughtered. Warkentin and the Mennonite settlers he helped organize would make their mark by escalating the transformation of the prairie into fields of grain.

Warkentin was one of many travelers on the prairie in the wake of the Civil War, willing and unwilling migrants to the American West. Native American pastoralists, disillusioned homesteaders, hungry squatters, and smooth railroad agents crisscrossed the prairie years before Warkentin appeared in the spring of 1871. Warkentin arrived in the United States in a semiofficial capacity, one of four scouts from southern Russia seeking agricultural lands to which the German Mennonites settled in southern Russia could immigrate. Fleeing compulsory military service, increased state control over education and governance, land shortages due to demographic expansion, and internal rifts between congregations, diaspora congregations of Mennonites eventually settled across the American West and in Canada, Mexico, Belize, Bolivia, Paraguay, Uruguay, Brazil, and Argentina.[2]

By the early twentieth century, the grain Mennonites cultivated in Kansas, known to settlers as "Turkey Red Wheat," had made the United States a breadbasket to the world. One agronomist described Turkey wheat as "an ancestral swamp to modern varieties," because all modern varieties have it in their lineage, including Norin 10, the famed semidwarf variety Cecil Salmon brought back from Japan in 1945 while traveling with MacArthur's army.[3] The semidwarf hybrids of wheat are credited with dramatically increasing world food production.[4]

Yet if Turkey Red Wheat provided the literal seed of the Green Revolution, it is virtually unknown beyond the heritage stories told about it, and these too require investigation, for there is a suspicious quality of stasis in what we know to be a history characterized by movement and instability. Fifty years after the introduction of Turkey wheat to the United States, Nikolai Vavilov celebrated its ancient lineage, using it to introduce his survey of world cereals, legumes, and flax. "The most common variety of winter wheat in the USA," he noted, "known under the name of 'Turkey' and lately occupying up to ten million hectares, or nearly one-third of the entire tilled area under wheat in that country, represents an ancient type of local Crimean wheat—'Krymka,' which was imported a few decades ago into the USA from Crimea."[5]

Vavilov's history prized stability, not flux, and he cited a story of origins that supported such a framework by focusing less on the ancient history of local Crimean varieties than on the people who introduced them to the Great Plains. Vavilov repeated the gospel of the USDA agronomist Mark Carleton, whose work with varieties introduced from southern Russia at the turn of the century increased their prominence and drew attention to their history. Carleton credited German Mennonites from this region, and Bernard Warkentin and Christian Krehbiel in particular, with introducing winter wheat to the Midwest in the 1870s. In 1914, fourteen years after his first trip to Russia, he provided a capsule history of their migration:

> The history of hard winter wheat in the United States is closely associated with the movement of Russian Mennonite immigrants to the middle Great Plains. These people originally went from West Prussia to southern Russia about 1770 because of certain land grants and civil privileges offered by the Government under Empress Catherine. One hundred years later their descendants desiring further advantages to be obtained in America emigrated to the middle Great Plains and settled principally in Kansas. . . . Each family brought over a bushel or more of Crimean wheat for seed, and from this seed was grown the first crop of Kansas hard winter wheat. Bernard Warkentin, a miller, who erected mills at Newton and Halstead, was chiefly instrumental in introducing the Turkey wheat, but in this pioneer movement of the Mennonites two other men were associated, Christian Krehbiel, first a

> farmer, but who later in 1886 erected a mill at Moundridge, and CB Schmidt, acting as immigration agent for the Santa Fe Railroad.[6]

His story of origins became the primary one, repeated thereafter by numerous historians of Mennonite heritage and agronomists seeking the origins of winter wheat.[7] Yet varieties by the same names were already under cultivation in the middle states.[8] Spring varieties of Black Sea wheat were in wide circulation. Agricultural periodicals reported Black Sea wheat for its drought hardiness and resistance to rust, with some reporting successful conversions of spring to winter wheat.[9]

Like any myth, Carleton's was a version of history, and perhaps this is better than no history at all: Carleton attempted to credit introducers of original material, not just the breeders who subsequently improved varieties. In this respect he was more even-handed than the agronomist who declared all nineteenth-century material an ancestral swamp to modern varieties. But in attempting to attribute credit to individual improvers, he compressed a history of multigenerational agricultural labor and biological exchange, such that all intervening history in between Vavilov's ancient local Crimean varieties and the Mennonites' conveyance of them to Kansas was lost.

Certainly the wheat the Mennonites cultivated was not the first in North America. European settlers on the Eastern Seaboard cultivated wheat they brought with them from Europe. The production of wheat increased consistently into the nineteenth century, mainly due to the westward expansion of cultivation, including drier and more hazardous areas. By the late eighteenth century, the Hessian fly kept wheat farmers on the move, fleeing the pest as well as the declining soil fertility of overused lands. Farmers demanded the opening of new territories in part so that they would have places to grow wheat.[10]

The Great Plains in particular was a hard place to grow wheat: windy and cold, with hard soil, the plants that could survive were susceptible to rusts, rots, scab, and bunt. The soft white winter and spring varieties that flourished on the Eastern Seaboard couldn't survive the harsh conditions of the plains. Nor were inherited methods of cultivation suitable. Whereas the total decline of wheat cultivation in the East had been staved off by the importation of European soil science and agricultural methods, these reforms failed in the plains, which were far more climatically dissimilar to northern and western Europe than the Eastern Seaboard states.[11]

Yet cultivation did flourish. Traditionally, historians have credited the mechanization of agriculture with rapid gains in productivity in the later nineteenth century. The steel plow broke the tough prairie cover. Seed drills enabled wheat to be sown more deeply than broadcasting by hand. McCormick's reaper saved the day, along with the Marsh harvester and wire and twine binders. Recent histories have also noted the extent of biological innovation: the transplantation and improvement of seeds that could survive the harsh conditions of the plains.[12] Chief among these were hard red spring and winter wheat, along with related varieties of durum wheat in the Pacific Northwest.

Farmers directed the improvement of wheat through pure-line selection and hybridization. In the northern plains, Red Fife became the most important variety shortly after its introduction in Wisconsin and Minnesota in 1860. The name "Fife" belonged to David Fife, a resident of Ontario who selected the plant. The seed that produced it was reportedly from a shipload of wheat from a port in Danzig, conveyed to a mill in Glasgow before reaching Fife in a small packet of winter wheat. Allegedly Fife's cow nearly sabotaged his contribution to the world food supply by grazing on the first planting: Fife's wife intervened.[13] And so Red Fife also became the parent of Marquis wheat, a hybrid variety that predominated in the early twentieth century.

In the southern plains, hard red winter wheat had the best odds of withstanding winterkilling and rust. Carleton traced the hard red winter wheat to the Russian steppe, "just north and east of the Black Sea and north of the Caucasus Mountains, including the governments of Taurida, Ekaterinoslav, Kharkof, and Stavropol, and the Don and Kuban territories."[14] Ecologically, the regions Carleton named composed the boundaries of the southwestern steppe: a region of fertile black soil, or chernozem, with a thick cover of wild grass and shrub similar to that of the midwestern American prairie. Traditionally used by Tatar nomads for pasture, the grasslands were ultimately plowed and sown for grain by Russian peasants and German immigrant farmers over the course of the nineteenth century.[15]

Politically, these regions included the territories of "New Russia" united by Catherine II's annexation of Taurida, territory formerly governed by the Crimean Khanate consisting of the Crimean Peninsula and the mainland between the lower Dnieper River and the Black Sea and the Sea of Azov. Control of Taurida, which gave Russia access to the Black Sea, was an outcome of successive Russo-Turkish wars and Russian imperial expansion in the eighteenth century.

Carleton, Vavilov, and Salmon all credited the Mennonites with the introduction of hard red winter wheat to North America. Carleton named Warkentin in particular as a progenitor. Verifying Carleton's story of origins requires following his trail to the Black Sea port of Odessa, down into the Crimean

Peninsula, back up to the Molochna River colony from which Bernard Warkentin emigrated, and again to the port of Berdiansk on the Sea of Azov.[16]

Bernard Warkentin was a young man when he arrived in Kansas in 1871, the twenty-five-year-old son of a prosperous miller in the Molochna River Mennonite settlement situated near modern-day Melitopol, about sixty-five kilometers east of the port of Berdiansk. He had arrived after a long voyage from Berdiansk to Hamburg and onward to Newfoundland and New York—a "dangerous and tiring affair" in which the ship "danced on the waves like a nutshell," he wrote to his friend Goerz. The fog lingered, delaying the ship's arrival into port. Then on the fifth of March, Warkentin wrote, "suddenly the fog was gone; and at 7:00 PM we saw a beautiful panorama, namely the continent of America. The place was called Staten Island."[17]

The next morning the ship docked at Hoboken, and the party traveled onward to the German Mission in New York City. "Here we found a life like we had seen it nowhere before," he reported to Goerz. "It is unbelievable that anyone could describe the tumult."

When Warkentin left Berdiansk, it was for a brief voyage. Without intending to, he became an exile. En route his fiancée died, and with no one to return to, he stayed the winter in Illinois. Soon letters from his family and Goerz described Russian authorities as hostile to those promoting emigration. Within a year, Warkentin had reconciled himself to remaining in America.[18]

Warkentin experienced the travails of any traveler in a new environment, but he relied on extensive networks of Mennonites already settled in the United States. While his letters were read avidly among congregations in and around Berdiansk, he was not so much a pioneer as a guest of the many Mennonites who had settled in Ohio, Indiana, and Illinois in the preceding decades. In Cleveland, he caught a fever and remained behind while his party traveled ahead. When he recovered he continued through Elkhart, Indiana, staying with Mennonites who had migrated from Prussia decades before. Soon Warkentin continued by steamboat down the Mississippi, which he thought looked like the Dnieper at home, only without the pretty islands and encumbered by shallow sandbanks. In Summerfield, Illinois, he met Mennonite elder Christian Krehbiel, who proved an enthusiastic advocate for immigration to western lands, touting their suitability for agriculture and cattle raising.[19] Together the two explored the Red River north and south of the Canadian border and followed the Missouri River from the Dakotas to Texas.

Krehbiel was some twenty years' Warkentin's senior, in age and arrival in the United States, and he assumed a leadership role in immigration. Like those of many Mennonite settlers, Krehbiel's life had been characterized by successive migrations. His family had first fled persecution in Switzerland before migrating to southern Germany and Bavaria. When his brother was drafted, the family moved again, this time to Buffalo, Cleveland, and Ashland, Ohio. With his family, Christian took what was then a thirty-five-day voyage from Le Havre. Devoured first by bedbugs, then by mosquitoes and German-speaking swindlers in pushing lousy hotels, Krehbiel set to work on a threshing machine at harvest in Ashland, stacking straw in the one-hundred-degree heat. He earned the admiration of his boss, especially in his prodigious production of manure made of lime and hay and the preparation of a clover patch for seeding, the art of which he had mastered in Germany. By 1861 Krehbiel had made his home in Summerfield, Illinois, where he struggled to make his one hundred acres turn a profit, mowing grass with a crooked scythe and cutting grain with a cradle.[20]

Twenty years later, Warkentin became a student of midwestern agriculture and lands. By the time he arrived in 1872, machines had largely displaced these rudimentary implements, with reapers, self-rakers, and self-binders enabling the cultivation of ever-greater swaths of land.[21] In their travels, Bernard Warkentin observed farms and their products with great interest, reporting on the cereals and vegetables cultivated, animal husbandry, and use of labor-saving machinery. Labor-saving machinery was of particular interest, perhaps not surprisingly given the scarcity of cheap labor in the United States. Wherever Warkentin went, he reported on the many different machines in use for harvesting. Warkentin was less impressed by the quality of the wheat being cultivated, noting that "in general the wheat in the United States is not the best," though that in Summerfield far exceeded the quality of that in Pennsylvania, Ohio, and Indiana. He noted that farmers sowed clover and timothy for feed with success.[22]

But Warkentin's was not simply a study tour. Railroad companies courted Warkentin and Krehbiel as agents of Russian Mennonite immigration. Seeking to fill their cars with agricultural products, they were eager to settle western lands quickly. Warkentin was convinced that if he accepted every invitation to view territory for settlement, he would never again stay still. By September, the two traveled west as guests of the Northern Pacific and Omaha Nebraska lines. Warkentin was hopeful about Oregon and Washington but noted with some anxiety the rapid increase of land prices almost everywhere.[23] Ultimately, he turned his attention to Kansas.

The choice of Kansas was not a foregone conclusion, and Warkentin was only the first of a number of delegates sent by Mennonite settlements and congregations to investigate settlement in the West. The Alexanderwohl congregation had first favored Palestine, then turned to Canada because of the British government's generous land grants and the terms of amnesty, guaranteed to be the same as those already extended to the Quakers. Meanwhile, Warkentin felt the country too cold, with fruit culture impossible, and the best lands already gone. He preferred Texas, while others felt the climate too hot and the people too strange. Ultimately, the scarcity of good lands and other factors dictated the choice of territory on the American side of the Red River Valley, in Minnesota; but when the deal went sour, the congregation settled for Kansas instead, on lands for which Warkentin and Krehbiel arranged the sale.

Accused of being boosters for Kansas, in fact both Krehbiel and Warkentin determined that in addition to the more southerly climate, Kansas's large tracts of contiguous land at affordable prices were crucial. These tracks were secured by the landholdings of railroads and the federal government's clearing of Indian reservations for land sales. The Atchison, Topeka, and Santa Fe Railroad Company offered the most favorable land prices. Krehbiel also reported that some 800,000 acres of Indian reserve land farther down the Neosho Valley had been "recently unsettled" and was being offered for sale as public land.[24]

There were conflicts over land. Near Parsons, Kansas, Krehbiel and his party found themselves threatened by a mob of squatters who refused the railroad company's intentions to seize and sell lands they had improved through settlement and cultivation. According to his autobiography, Krehbiel told their would-be attackers that his community would never contribute to others' losing their land, and that as farmers themselves, they would side with their own kind, not the railroad company. Whether or not this was true in the immediate context, public lands were made so through government policies of forcible unsettling of American Indian inhabitants. Krehbiel advised readers that investment was imprudent until these ownership questions were settled.[25]

Warkentin bristled at the competing intelligence of the congregational delegations and felt keenly aware of his youth and inexperience, but his early settlement and connections to the railroad companies made him the point person when it came to the effective coordination of movements. When the first large families of settlers of Peter and Jacob Funk arrived in early 1873, Krehbiel and Warkentin traveled with them at the behest of the Santa Fe Railroad Company along its new line, using the car as a hotel while examining nearby lands. Krehbiel took the lead on negotiations with the railroads over the terms of settlement.

Wherever the party went, they dug soil three feet deep, studied the railroad company's maps and appraisals, and offered the advance party its choice of

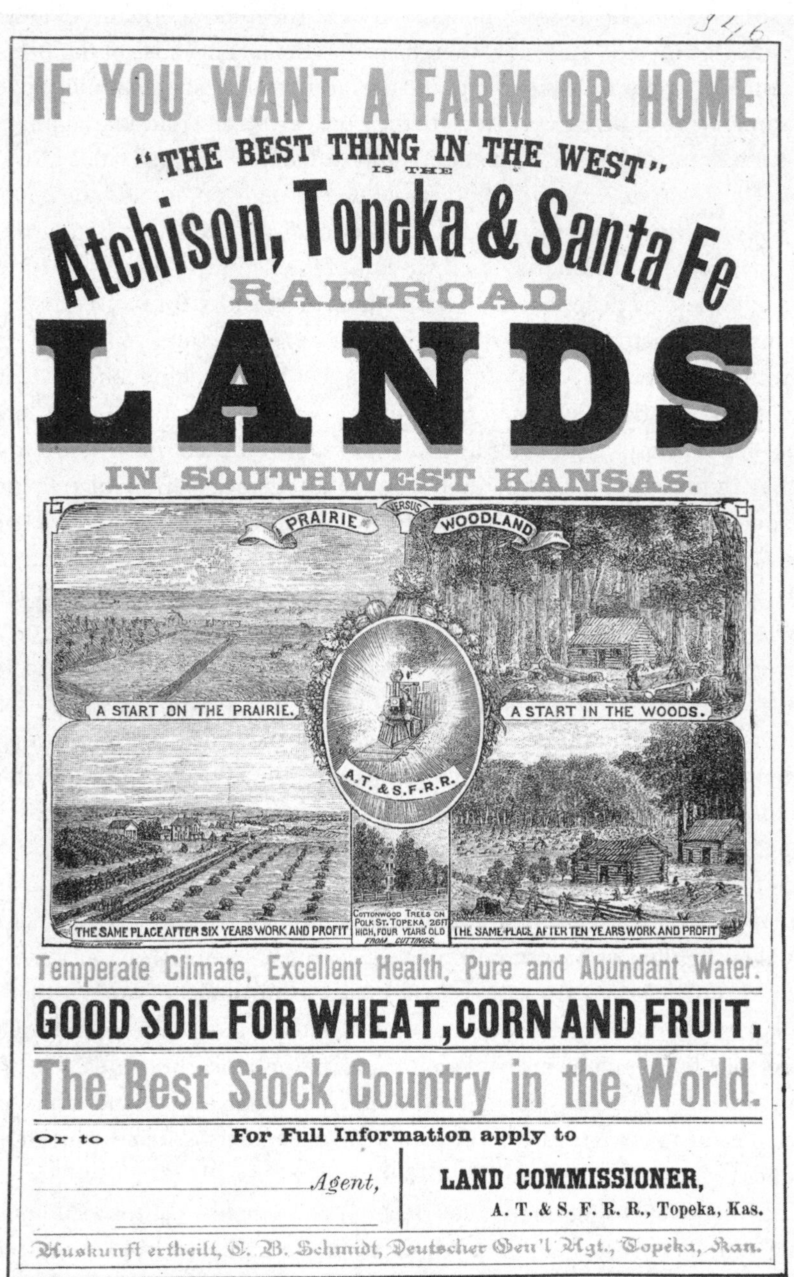

Figure 4.1. Broadside advertising lands for sale by the Atchison, Topeka, and Santa Fe Railroad Company, with German-language instructions to receive information from German general agent C. B. Schmidt.

land near waterways, woodlands, and flat prairie. The traveling party was rich, and Krehbiel and his companions succeeded by bulk purchase. Krehbiel arranged to reserve all lands in the area, along with water rights and a mill site, which was purchased by Warkentin and John Wiebe. When Krehbiel secured the land at $2.50 per acre, the Funks produced $50,000 in cash from their baggage and paid in full. Thereafter an uneasy Krehbiel noticed the greedy and menacing glances of many fellow travelers toward their luggage and persons.

For their part, the agents of the railroad company redoubled the official commitment to courting Mennonite settlers, bringing their own experience to bear on the project of settlement. A. S. Johnson, the party's guide through Kansas, was the son of missionaries, then reputed to be the first white child born in Kansas. C. B. Schmidt, who in late life was anointed the "Moses of the Mennonites" for his encouragement of immigration, was himself born in Germany and immigrated to Illinois at the age of fifteen. Schmidt translated Krehbiel's German to English for all contractual dealings with the railroad company. Later he traveled to Prussia and onward to Russia in an attempt to sway those who opposed emigration, or who favored Manitoba over Kansas as a destination. He wrote to Johnson of the hospitality and prosperity of all the Mennonite farms he visited.[26]

For all their diplomacy, however, the railroad agents wielded enormous power in shaping the western landscape. Patterns of settlement were ultimately determined by the holdings of the railroad companies, not individual preference: the grid of corporate landholdings arranged settlers on the plains. Between 1872 and 1875, Krehbiel went west repeatedly. All the lands around Marion Center and Newton appeared viable to him. Shortly after his initial trip west of Topeka, he returned to buy up lands near Halstead he had noticed from the train.

By the spring of 1874, Warkentin had partnered with Krehbiel and the brothers Funk to form a Board of Guardians in Halstead, vested with the responsibility of supporting Mennonite immigration to the United States. They made arrangements with the Red Star Line and the Pennsylvania Central and Santa Fe Railroads for the passage of immigrants, crisscrossing the country meeting and accompanying hundreds of families from the largest congregations in southern Russia, including the substantial Kleine Gemeinde and Alexanderwohl congregations.

When the first hundreds of families arrived, it was with the bare necessities: no more than eighty pounds of luggage each, Warkentin having instructed them to bring coats and linens but leave cutlery and kitchen goods behind. Warkentin's friend Goerz was among the early families to cross, though he found himself delayed in Hamburg by errors in his visa. When it came time to board the ship, he unexpectedly encountered his own parents, who had traveled to the United States months in advance only to return almost immediately. As the final all-aboard sounded, they could only warn him in passing that he was bound for a wicked land. He boarded the ship with tears in his eyes.

Certainly there was no honeymoon. Goerz's parents wouldn't have known, but beginning in 1873, swarms of locusts and grasshoppers overtook the West. By the time Krehbiel moved to Halstead in 1875 with the thousands of settlers arriving daily via the Red Star Line and the Santa Fe, the grasshoppers were so thick the fields were bare and black, and railroad cars slipped on the tracks. In farm wagons instead of buggies drawn by fleets of ponies, Krehbiel remembered, "all illusions vanished. Only courage and faith in God remained."

Faced with such a prophecy, why did Goerz and others go? When they set their sights on the United States, it was not so much a fantasy of opportunity as one of distress. By the 1870s, Alexander II's edicts threatened to revoke the Mennonites' amnesty from military service, as well as their privileges of autonomous local government and independent schooling. Recent droughts had threatened many otherwise prosperous farms. Moreover, demographic growth within the Mennonite community had exacerbated inequality between landowners and landless Mennonites, and to a lesser extent between landless Mennonites and Russian laborers. Conflicts between conservative religious leaders advocating total separation from the world and progressives seeking commercial prosperity exacerbated rifts among congregations. While many congregations sent delegates to Saint Petersburg to negotiate the terms of their settlement, their returns were not promising. As conditions deteriorated, many Mennonite congregations looked for new lands to recreate the prosperity they had found during the previous century in south Russia.[27]

In Russia, the Mennonites were celebrated agriculturalists, but we have to look beyond almost clichéd celebrations of their prosperity for explanations as to how and why they succeeded. The commercialization of grain production in the steppe occurred where property was organized along individual and cooperative, if not capitalistic, lines, with robust institutions of credit and entitlements to profit from surplus production.[28] Mennonites also benefited from state grants of land and privileges, which enabled them to build up surplus capital for investment in commercial agriculture.

In accepting Catherine II's invitation to settle lands in southern Russia, the Mennonites, with the other foreign colonists, became instruments of Russian imperial expansion. The Mennonites were a diaspora community of Anabaptist agriculturalists, previously settled in West Prussia before excessive taxation and compulsory military service prompted waves of migration. Catherine, a German by birth, courted European colonists.

By settling Russian land annexed in the context of successive Russo-Turkish Wars in the late eighteenth century, foreign colonists facilitated the consolidation of the southwestern frontier and eased access to the Black Sea ports to the south. Catherine refigured the southern steppe and the Crimean Peninsula as "New Russia," pursuing a policy of aggressive colonization.[29] Nor were Mennonites the only settlers in the region. The Russo-Turkish wars instigated movements of Greek, Armenian, and Jewish merchants to the Black Sea trading ports in the territory of the former Crimean khanate. The rural steppe north of Crimea was settled by Crimean and Nogai Tatars. Mennonite settlements also abutted large communities of Catholics and Lutherans. Other religious minorities, including Jews and Dukhobors, a Russian spiritualist Christian sect, settled in nearby villages, as did Russian Orthodox state peasants. Beginning in the 1840s, communities of Hutterites often joined Mennonite settlements.

In exchange for settling the frontier, Mennonites and other foreign colonists received privileges and grants that gave them more power and wealth than Russian state peasants. They received control over local governance and education, amnesty from military service, and preferential grants of state land. Much of the land was expropriated from Nogai Tatars, seminomadic pastoralists forcibly resettled from areas north of the Sea of Azov to the Caucasus and back again, as ongoing conflicts with Turks risked alliances between the Turks and the Nogai.

Catherine's overtures to foreign colonists were of a piece with her broader practices of resettlement. According to her 1763 manifesto, foreign nationals settling vacant lands in the southern steppe would receive transit funds, no-interest construction loans, duty-free import of personal effects and sale goods up to three hundred rubles, ten years' exemption from import duties, thirty years' exemption from "all taxes and obligations," and permanent exemption from military or civil service. They could also own serfs. The Mennonites received the strongest of all protections from the state in terms of land control and autonomous government, formalized in the Privilegium granted by Czar Paul in 1800.[30]

The Mennonites realized the intentions of the Privilegium in acting as model colonists and commercial farmers, often generating friction with Russian peasants and Tatar populations.[31] The first Germans, Mennonites from West Prussia, had arrived in 1789 and settled in Taurida on the banks of the Chortitza River. Beginning in 1805, some twenty-four groups of Mennonites settled along the Molochna River, just north of what is now Melitopol, Ukraine. Soon the Molochna Mennonite settlements with their picturesque villages became a stop for royals en route to their palaces in Crimea. Presented as the fruit of Potemkin's wide-ranging program of rural reconstruction, Mennonite farms benefited from the Russian regional government's efforts to modernize agricultural practice through the foundation of progressive agricultural societies.[32]

In Molochna, the leader of the progressive agriculturalists and the president of the Agricultural Society was Johann Cornies, a Prussian-born emigrant with entrepreneurial flair. Russian state officials mindful of Cornies's success in commercial husbandry enlisted him in the foundation and administration of agricultural societies. In 1817, Cornies had acted as a member of a settlement commission organizing the distribution of land to new Mennonite immigrants. In 1824, he directed the Association to Improve Sheep Raising. By 1830, he led the regional Forestry Society, and five years later, the new Agricultural Society. In this capacity, he directed numerous reforms to adapt arable husbandry to the water-scarce environment of the steppes. These included dams and irrigation, the introduction of a four-field crop rotation system, the use of horse manure as fertilizer, deep plowing of fallow land, and the planting of orchards, forest-tree plantations, and mulberries for silk production. (The orchards in particular irritated pastoralists, who complained that they could no longer spot their herds across a field.) Under Cornies's leadership, yields of winter and spring wheat increased dramatically.[33]

By the late 1830s, the Ministry of State Domains, charged by Nicholas I with reforming the Russian peasantry, had tasked Cornies with modernizing his neighbors, including the Nogai, whose mobility and use of lands for grazing presented a particular problem for agricultural settlement. Meanwhile, the state pursued an aggressive campaign to sedentarize nomadic pastoralists. An 1816 edict denied travel passes to Nogai men who did not sow two chetverts (about twelve bushels) of grain, preventing them from traveling to Crimea to work as herdsmen.[34]

Cornies, viewing Mennonite settlements as a model for commercial agricultural production, set about reconstructing Nogai, Jewish, Hutterite, and Russian peasant villages along the same lines. Mennonite settlements provided sites for experimentation in plant varieties, agronomic methods, and social control. Having previously engaged in projects to improve Nogai sheep, in the early 1830s Cornies set about sedentarizing Nogai pastoralists in the village of Akkerman, which abutted Cornies's own estate on the Iushanlee River. Cornies plotted Akkerman according to precise instructions. These reflected the Mennonite ban on proselytizing by encouraging adherence to Muslim law. In matters relating to husbandry and rural economy, however, they were less catholic. Thirty-five articles defined village construction and administration, from gables to gates, with paint, lot width, and tree rows all precisely indicated. Four-field cropping and manuring after the Mennonite model were mandatory. The perfect village, as it turned out, was a Mennonite village.[35]

Cornies's style of improvement provided a narrative of commercialized agriculture as civilization that served the interests of the Russian state. He disdained Nogai agricultural practice, which he regarded as a "lottery."[36] Cornies complained that Nogai farmers simply dropped seeds in the ground and watched them grow. In reality, the Nogai, like other nomads in the steppes, practiced a shifting agriculture, oriented toward temporary settlement and subsistence production, cultivating small amounts of grain for local use. While Cornies's reforms helped organize and increase commercial agricultural production, they also contributed to stereotypes of Slavic peasants as backward and resistant to change, and of Tatar pastoralists as nomads in need of civilization, rendered synonymous with arable husbandry and sedentary life.

This pejorative characterization of nomadism implied that pastoralists had never been sedentary, and were indeed incapable of settled society. But only in the stuff of myth is there such a thing as a pure nomad. All communities exist on a continuum between sedentarism and nomadism, and all nomadic communities depend on sedentary societies for survival. Ironically, the purest form of nomadism may be novel rather than traditional, the result of forcible displacement by the state or foreign colonists.[37]

In fact, the long-standing presence of Tatar settlers in the steppe, evidenced in part by the repeated contracts between Molochna Mennonites and Nogai Tatars, indicates persistent and evolving patterns of land use, including arable husbandry. Arable husbandry was a minor part of the Nogai economy, which reserved scarce water resources for livestock. While Cornies enacted the will of the state in sedentarizing Nogai populations, sheep husbandry remained more profitable for the Nogai, especially as the Mennonites encouraged their further specialization in pastoral husbandry. The contraction of the international wool industry in the 1830s threatened Nogai livelihoods. When they turned to arable agriculture out of necessity, only the most marginal lands remained.[38]

Yet it is as incorrect to characterize the Nogai as averse to agriculture as it is to regard their pastoralism as static and timeless. The Nogai attracted an undue portion of criticism from state authorities because of the challenge they presented to Russian settlement of the steppes for commercial agriculture, but this did not in itself indicate the absence of longer standing agricultural practices. Even Cornies never denied that the Nogai farmed. He simply said they were bad farmers.

In reality, the Nogai farmed for different purposes, oriented toward subsistence and supplement rather than surplus and market. Accounts of Cornies's leadership and agricultural innovation may tempt us to think that Mennonites adapted to their environment better than other settlers, but the reverse may be true. The Nogai pastoralists who settled on the grasslands for generations before the foreign colonists arrived worked in equilibrium with the dry grasslands, cultivating grain only inasmuch as required for local consumption. Newly arrived Russian peasants, too, quickly abandoned intensive agriculture in favor of pastoralism, gardening, and long fallow agriculture optimal for subsistence production. The Mennonites, in contrast, were stubborn in imposing agricultural techniques and implements brought from afar. They were determined, even in the face of repeated crop failures and shortages, to make the land pay. Ultimately, they adapted their inputs and techniques to defy environmental limits, including fertilizers, irrigation, crop rotation, and deep plowing.[39]

Mennonite agricultural knowledge was also not as self-contained and sui generis as it appeared. Cornies and other Mennonites studied knowledge and techniques from Germany and elsewhere in Europe, often maintaining considerable libraries on agronomy and experimentation as they attempted to adapt to the dry climate of the steppes.[40] But it is easy to overemphasize the European aspects of Mennonite husbandry by focusing exclusively on their libraries and self-identification as German agriculturalists. The Mennonites experimented with many innovations drawn from European practice, yet the seeds they cultivated were from territories surrounding the Crimean Peninsula settled by Nogai and Crimean Tatar populations.

Even before Mennonite settlement, access to local grain was a boon to Russia in successive campaigns against the Ottoman Empire. Provisioning troops on the southern border headed to Crimea posed a logistical challenge. Seventeenth-century military campaigns in the south relied on moving grain from northern court lands and special provisioning from the Belgorod, Kursk, Sevsk, and Don River regions. Smallholder servicemen formerly engaged as local militia produced grain for infantry on the southern frontier. By the end of the seventeenth century, this southern supply system supported Russia's expansion beyond Russia's southern border, into the steppe and Crimean Peninsula. Tatars too were compelled to supply Russian troops.[41]

Ultimately, the rise of New Russia as a grain exporter was a geopolitical achievement as much as an agronomic one. While it is tempting to credit Cornies with the boom in commercial wheat production, his ambition played a fairly minor part in the story. The 1774 Treaty of Küçük Kaynarca between Russia and the Ottoman Empire secured Russian rights of commerce in Black Sea ports and rendered the Crimean khanate an independent state, shifting the balance of power in the Black Sea region from the Ottoman Empire to Russia. Catherine's annexation of the Crimea a decade later secured Russian dominance of the region, refigured as New Russia.[42] By 1806, the Turks allowed French and Dutch ships to pass the Bosporus, and large amounts of Russian grain began to reach western markets, compensating for disruptions in production generated by the French Revolution and the Napoleonic Wars. Access to Black Sea ports and expansion of grain cultivation in the southern steppe benefited European markets, and also Russian ones.[43]

Far from simply an "Ottoman lake" secured by the conquest of Kaffa in 1475, the Black Sea became a theater of exchange linking Ottoman and Russian empires. Until the Treaty of Küçük Kaynarca rendered the Crimean khanate independent of Ottoman control, the Black Sea connected Istanbul to the northern reaches of its empire, linking the ports at Ochakov and Kinburun on the Dnieper River, Kili on the Danube, Akkerman and Bender on the Dniester, Azov on the Don, and Kerch and Taman linking the Azov and Black Seas.[44] Moldavia, Wallachia, and the Crimea provisioned Istanbul with grain as well as livestock, metals, and wood. With the cession of the ports of Kinburun and Kerch to the Russians, Russians traded freely in the Ottoman Empire. The embattled Crimean khanate, weakened by Zaporozhian and Don Cossack raids, lost the power it had enjoyed during Ottoman rule.[45] Within this reconfigured trading region and under the auspices of Russian-Ottoman commercial agreements, Ottoman *reaya* nevertheless continued to cultivate their lands on Black Sea ports. Near Kinburun, *reaya* continued to export grain via the Russian-controlled ports.[46]

In the first half of the nineteenth century, New Russia became the principal source of grain to western Europe through the Black Sea and the Mediterranean, becoming the principal exporter in a far-flung deposit trade moving grain from Black Sea ports to Italian and Turkish ports in the Mediterranean. From there it went onward to a series of intermediate ports, where it might be held for as many as seven years, depending on international market prices. In 1836, the port at Berdiansk opened, dramatically shortening the transit of grain to market for farmers in Molochna (some sixty-five kilometers east) and adjoining regions. British repeal of the Corn Laws in 1846 stimulated international trade by removing protective tariffs favoring British producers, marking the end of a more gradual shift from protectionism to liberalization in international trade. Meanwhile, the decline of the wool industry hastened the transition to wheat culture, incidentally incurring economic ruin for Nogai Tatars, who had specialized in sheep husbandry in part because of Mennonite investment.[47]

The expansion of commercial cultivation depended on local seed stocks. A variety of wheats flourished in the black soil of New Russia, including hard and soft winter and spring wheats, with spring wheats often preferred in the colder climates of the northern and eastern steppe. While soft wheats predominated in north European markets for bread and biscuits, durum wheat suited to pasta sold well in the Mediterranean and Italy in particular. Girka wheat, a soft winter and spring variety introduced in the 1840s, predominated in the trade with the British into the 1850s. Among the durum wheats, the most commonly sowed variety was Arnautka, noted for its high gluten content and reported by one German traveler to a Tatar villager as "generally beautifully large and of light coloured grain." Others referred to the same varieties as Turka, Beloturka, Kubanka, Krasnoturkaia, or simply Russkaia pshenitsa (Russian wheat).[48]

Red varieties were characterized by drought hardiness, which made them well suited to the dry steppe. State harvest reports explicitly note "Red Wheat" growing in Molochna as early as 1812, in addition to the general categories of winter and spring wheat listed in earlier reports.[49] David Moon notes that by the late 1830s, Peter Köppen recorded farmers replacing Arnautka with hard red winter wheat from the Crimean Peninsula, which was then in demand at the ports of Berdiansk and Mariopol. Red Sandomirka wheat, a soft variety similar to Girka, was especially popular. By the 1840s, the Ministry of State Domains was conducting extensive experiments to ascertain the best crop varieties for yield and drought resistance.[50] While the drought tolerance of red wheats recommended them in the dry and comparatively warm peninsula, by the 1870s, Mennonites in the colder mainland region had to devise a technique for protecting young shoots from the winter frost by harrowing soil over them. Moon concludes that by the turn of the century, hard, red winter wheat of Crimean origin predominated in the steppe region north of the Black Sea and the Sea of Azov, and in the near north Caucasus.[51]

Tracing the exact path of particular varieties of seed grain to or from the Crimean Peninsula north through the steppe presents a number of difficulties. In general, macroeconomic histories of trade relations emphasize exports rather than the social worlds and complex practices of production they distill. Although commercial data for the Black Sea region clearly indicates the expansion of international trade and the disruptions of war, it provides few clues as to the origin of particular varieties of seed and other agricultural inputs. Nevertheless, the rapid commercialization of the wheat trade in the nineteenth century, and the concomitant expansion of mercantile networks in the Black Sea region, suggests mobility of grain varieties both backward and forward from the Black Sea ports to the farms in the steppes. Mercantile and trade networks can move an array of materials in multiple directions, even if only that which is monetized is documented.

Merchants, carters, and shippers all played a role in moving grain to market, and they may have also played a vital role in moving knowledge and material to centers of production. While new Russian ports at Mariopol (1779) and Berdiansk (1834) brought wool and grain to international markets, bringing products to port continued to rely on seasonal feats of inland transportation, witnessed by the US consul as a great caravan arriving in Odessa from some five hundred miles away in the steppe. Horse- and oxen-drawn carts departed in late spring and summer when the roads were dry and hard again, and then again in early fall. Each cart could carry five to six chetverts of hard wheat or six to eight of soft, such that in 1845 some 200,000 carts went to market. Foreign colonists and Russian landlords typically accompanied their wheat to market, often acting as merchants themselves. Carters, or chumaks, including Cossack and peasant populations, did the heavy lifting.[52] By the 1840s, Nogai Tatars also worked carting grain to market. Each of these populations may have played a role in selecting and returning seed grain to settlements in the north.

The government also often provided seed grain to farmers in years of depression and to compete with the increase of American grain on the market in the 1840s. The Imperial Society of Rural Economy, based in Moscow, sponsored research into varieties suited for steppe cultivation. It enlisted the US consul to obtain varieties of wheat native to Michigan and Japan reported to be growing in the prairie.[53] Cornies, too, became a state agent for experimentation with new crop varieties as he supported broader directives of improvement of plants, animals, and people. The botanist Christian von Steven, who had directed silk production in the Caucasus and Crimea before being promoted to general inspector of New Russian Agriculture in the Ministry of State Domains, sent a wide variety of plants for trials drawing on his extensive knowledge of Crimean flora.[54]

More essential than tracing the immediate introduction of hard red winter wheat to the steppe, or to Kansas for that matter, is acknowledging that the Crimean varieties conveyed to the steppe drew on generations of agricultural practice, at least partially of peoples cast as indolent and nomadic by state policies of removal. Into the 1830s, travelers reported the Turkish and Tatar populations in Crimea little interested in commercial agriculture, noting that the Russian government was allowing them to move to Turkish territory because they were "indolent in character."[55]

Although Cornies and others may have favored input-intensive agriculture oriented toward large-scale production, when it came to finding seed grain best adapted to local conditions, they relied on seed preserved by those who had been farming in the region the longest. Success in expanding production thus depended on a longer standing and more modest style of cultivation.

The ascent of Russian authority and commercial grain production in the region should not obscure prior agricultural practice in the region. The dispersed lands of the Golden Horde, controlled by the descendants of Genghis Khan's army, consisted of numerous resettlements of Tatar populations. Many of the Nogai Tatars in the Molochna region had migrated from the steppe border of the northern Caucasus in the sixteenth century, when the Muscovite state conquered the khanate of Astrakhan. They became informal subjects of the Crimean khanate, the Turkic-Mongol state originating in the early fifteenth century, which administered regional trade and tribute for the Ottomans until 1784.[56] The khanate as a whole encompassed Black Sea trading zones populated by Armenian and Greek merchants, as well as numerous pastoralists and nomads north to the Danube and the Balkans. Nogai pastoralists acted as a buffer against the Russians to the north and trading partners with their neighbors to the south.

The Crimean khanate was not a nomadic but a sedentary, agricultural economy. In the new Russian state, Tatar nobles were assimilated or marginalized, but formerly they managed large landholdings similar to those of Russian nobles in the north. Tatar peasants who worked the land owed nobles one-tenth the value of the grain harvest, one-twentieth of livestock products, and a variable corvée (unpaid labor).[57] Ottoman agriculture on the whole revolved around peasant smallholders (*reaya*) on state-owned land (*miri*) cultivating primarily bread wheat, a tithe of which was collected by cavalry (*sipahi*) who resided with the peasants and acted as functionaries of the state.[58]

Although Catherine regarded Tatars as an inconvenient obstacle to Black Sea Ports, treaties securing Russian control of the region specified that they would retain their rights to land. Yet she and her regional governors struggled to understand the structure of Crimean landownership, which consisted of a variety of arrangements between the khanate, *mirza* nobility, and peasants.

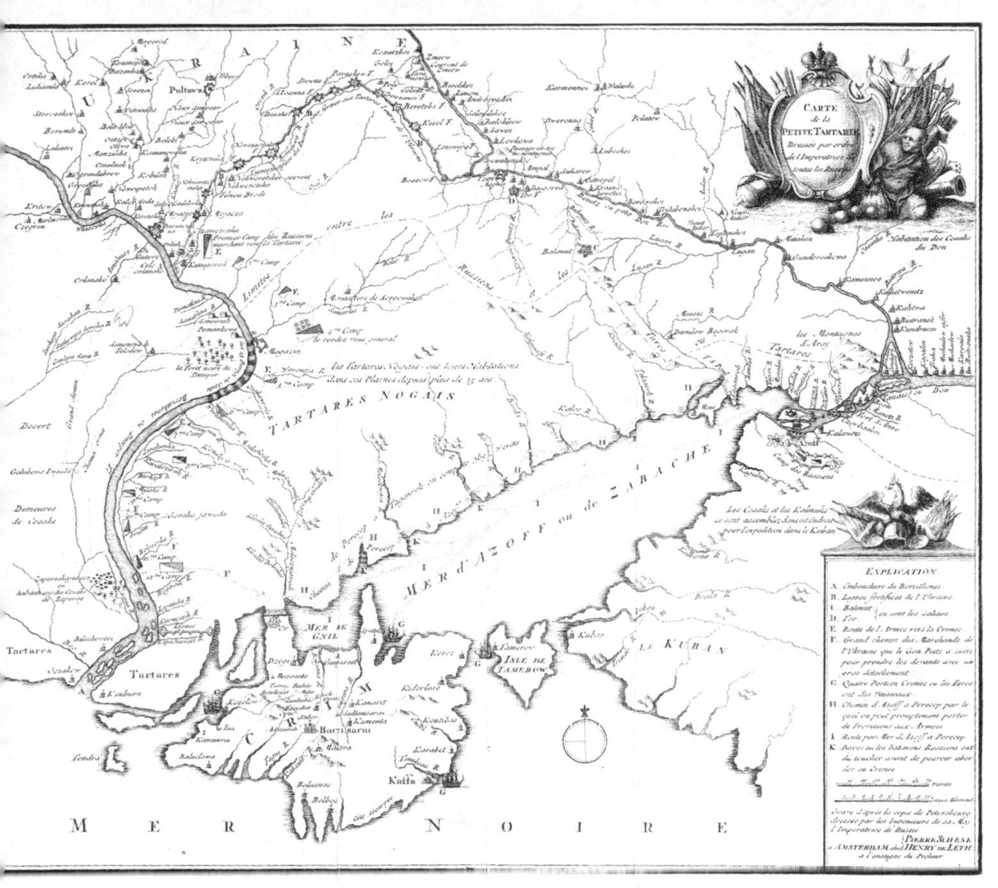

Figure 4.2. Circa 1785, Carte de la Petite Tartarie, Amsterdam, Pierre Schenck, Henry de Leth. Based on an earlier Russian map, this production of the Dutch cartographer Hendrik de Leth demonstrates the settlement of Crimean and Nogai Tatars north of the peninsula during the Russo-Turkish Wars of the eighteenth century, including Russian territorial boundaries and troop movements. Catherine the Great's annexation of the peninsula and northern steppe in 1783 established the Russian province of Taurida. Lada-Mocarski Map Collection, Yale University.

Grants from the khan pertaining to settled lands included freehold *mevat* grants and *timar*-style grants defined by the right to collect revenue. Over time many *mirzas* converted the latter into *beyliks*, hereditary lands outside the control of the khan. On the whole, Tatar peasants also enjoyed greater rights to their land than the Russian state equivalents, and so they resisted designation as Russian state peasants in the wake of annexation. As governors struggled to define a policy regarding land rights, frictions between Tatar and Russian landowners increased. The latter claimed that Tatars harvested grain, gathered

Figure 4.3. Map from 1822 of the province of Taurida extracted from an atlas compiled by Russian military cartographer Colonel V. P. Piadyshev for the Geographic Atlas of the Russian Empire, the Kingdom of Poland, and the Grand Duchy of Finland. The map shows roads, postal stations, customs offices, and boundaries of provinces and districts with distances measured in versts. World Digital Library, Library of Congress.

fruit, fished, and pastured animals on Russian estates without any compensation. Tatar settlers also lacked interest in commercial grain cultivation, plowing only a small percentage of settled lands.[59] They nevertheless had a long-standing agricultural practice.

Beyond the immediate legacy and geography of the Crimean khanate, predominately sedentary, agricultural economies had characterized the black soil region of the Volga River for centuries. Travel narratives record barley, wheat, and millet cultivation in the pre-Islamic Bulgar khanate, with single share, heavy metal plows used to break stiff soils. Linguistic evidence of Tatar folk traditions documents the festival of the plow (*sabantui*). Records of taxation provide another source of evidence for agricultural practice, with Volga cultivators in the Bulgar state obliged to pay rulers one sable pelt per household per year. The Kazan khanate (1438–1552) had extensive land taxes, indicating a sedentary agricultural economy with roots in its Bulgar past. These specifications also indicate extensive commercial ties between regions. This far-flung trading network in turn suggests a broad movement of seeds and plants from an early period, before the Mongol conquest and through the dispersal of the Golden Horde lands.[60] Foreign colonists and Russian settlers drew on this long tradition of agricultural practice as they sought varieties for commercial grain production.

The Black Sea trade integrated diverse and often warring populations into *long durée* networks of exchange, including sparse rural sedentary and nomadic populations in the steppe frontier. These networks trafficked in grain, goods, and people. The Crimean khanate controlled the peninsula, while in the steppe to the north, Tatar pastoralists in the Volga-Don-Kuban region and Bucak steppe abutted sedentary Zaporozhian Cossacks on the Dnieper.

Slave raiding was an entrenched part of the economy and it functioned in conjunction with broader trade. The center of the slave trade remained in Kaffa (Fedosia) on the southeast of the Crimean Peninsula, moving Slavic and Tatar slaves through the Ottoman Empire.[61] But the slave trade flourished at other ports, too. At Akkerman, Slavic slaves were conveyed to Ottoman markets along with grain, silk, spices, wine, wax, and hides.[62]

Vavilov may have been right when he said Turkey wheat was of ancient Crimean origin. Demosthenes (384–322 BC) reported that Greek cities depended on Black Sea land for grain. In the thirteenth century, Venetian and Genoese merchants carried grain from the Black Sea and Crimea to Constantinople and Italy.[63] If the Turkish conquest of Kaffa in 1475 inhibited western trade, wheat did not grow itself for the next three centuries, nor had it during the steppe's settlement by Scythian nomads in the millennia prior. The success of cultivators in the northern steppe of Taurida had something to do with Mennonite agronomy, but also to the Black Sea's continuity as a trading region and its history as a borderland between Russian and Turkish expansion governed for three centuries by the Crimean khanate. When yield increased, in New Russia, and again in the United States, it was less attributable to introduction of the seed itself than commercial imperatives to grow more grain.

By the time Cornies died in 1848, the transition from mixed agriculture to pastoralism to commercial grain production in the southern steppe was well underway. While sheep rearing remained important until the end of the 1850s, especially for landless settlers who grazed on land leased from the Nogai, the Crimean War exacerbated the increasing tensions between wealthy landowners and landless peasants by increasing food prices and stimulating a demand for grain. Most remaining Nogai Tatars left in 1859 and 1860 for the Ottoman Empire, their remaining land used to resettle Russian and Bulgarian colonists. While the Mennonites held their standing in New Russia, over half of their settlers had no land. Cornies had envisioned a sharecropping system in which commercial agriculture would thrive, but only those with land could prosper.

After Cornies died, his son-in-law Philip Wiebe assumed leadership of the Agricultural Society and control of the experimental farm at Iushanlee.

Although many immigrated to Kansas and Nebraska in the 1870s, many remained behind or returned. Bernard Warkentin retained close contacts with his family and friends in Molochna and Crimea, including his brother-in-law Johann Wiebe, who forwarded several shipments of seeds at Warkentin's request in the decades after migration. In advance of Mark Carleton's trip to Russia, Warkentin wrote to Wiebe requesting seed grain and help searching for new varieties.[64]

Warkentin thought it best to go to the source. "It has always been my idea that if it [wheat] could be found in the Crimea, from where it was originally brought here to Kansas by Mennonites, which settled in 1873 in Marion County," he wrote to Carleton, "that it would be preferable."[65] At the time of Carleton's departure in the spring of 1900, Warkentin had already contracted both Wiebe and his other brother-in-law in the Crimea, Jackob Enns, to order a hundred bushels of wheat. He gave Carleton detailed directions and shipping instructions, suggesting that it was not the first such shipment. He also explicitly mentioned a shipment made five years before by Wiebe, suggesting that new shipments of seed were not uncommon in the 1870s and 1880s.[66]

While Warkentin repeatedly stressed the necessity of obtaining pure and clean Turkey seed, it is likely that in practice seed was mixed with a variety of types in cultivation. Carleton helped facilitate a private purchase of more wheat from the region for Warkentin's milling company. When Carleton's first shipment of wheat arrived in November 1900, it was of insufficient quantity to seed alone. Unable to wait any longer, farmers mixed it with existing stocks. Warkentin nevertheless renewed his efforts to obtain several thousand bushels of pure seed for the next year's seeding.

Once Carleton had traveled extensively through Russia, he had his own advice for Warkentin as to where the best seed could be gotten. Central Crimea near Kurman-Kemelchi provided the best stock, as well as Warkentin's homeland near the Molochna River. In 1900, Carleton introduced Kharkov wheat, named for the area on the steppes of Starobyelsk where he collected, along with more Turkey wheat from Molochna. Kharkov rivaled Turkey for hardiness and productivity, and Carleton's introductions of durum wheat made major contributions to cultivation in the Northwest. While Carleton credited Warkentin for the introduction of Turkey wheat in his subsequent publications, the locus of expertise had shifted.

By then, the title of Turkey wheat was linked to those of the Mennonites. There was no further discussion of its origins beyond the folk name attached to it: either a geographic or ethnographic designation, alluding vaguely to the Crimea's past as a vassal state of the Ottoman Empire. The name, innocent

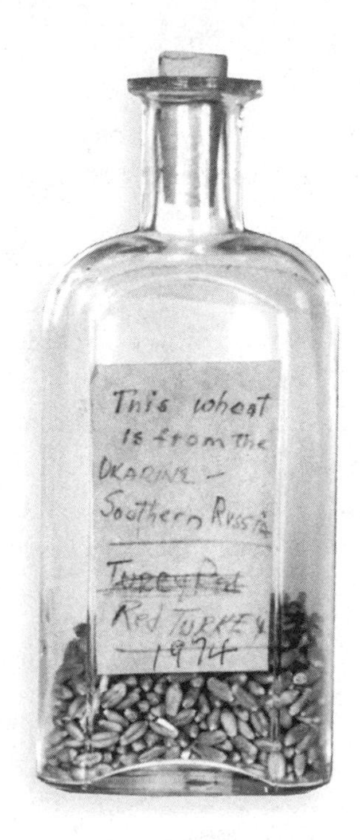

Figure 4.4. "This wheat is from the Ukraine—Southern Russia / Turkey Red Red Turkey—1974." Charles Goebbel, of Burlingame, Kansas, purchased this bottle containing Turkey Red Wheat seed in 1974, when residents of Goessel, Kansas, sold bottles to commemorate the centennial anniversary of the introduction of the variety in Kansas. Kansas State Historical Society.

enough on the face of it, obviated the intervening Russo-Turkish Wars, Russian imperial overtures to foreign colonists, the concomitant displacement of Tatar populations, and campaigns to sedentarize pastoralists on marginal lands. Meanwhile, the mythology of nomadism obscures the forced conditions of mobility, and the policies of removal that stripped both Tatar pastoralists and Plains Indians of their land. These received histories obscure policies of removal and ignore the diversity of societies from which pastoralists originated.

When Carleton credited Bernard Warkentin and Mennonite settlers with the introduction of Turkey wheat to North America, he generously substituted a myth of cultural heritage for a self-serving story of scientific advance. But heritage, like innovation, is a static and proprietary category, denying the social and material exchanges that produce knowledge. As a community bound by ideology and religious conviction, Mennonites valued their distinctiveness, and their identity stories shored up narratives of innovation and progress that celebrated autonomy, self-reliance, and entrepreneurialism.

But the Mennonites were not self-reliant at all. They were colonists, constantly on the move, dependent on connections with other communities for trade, labor, and subsistence. Those exchanges were crucial to corporate survival, but the will of the community to remain separate obscured them.

Sources of knowledge vanished at the points of exchange, perhaps especially when social and economic relations were hostile or fraught, as with the Tatar populations among which foreign colonists settled. This erasure produced varied myths of independence: the stuff of Mennonite prosperity or nomadic defiance.

The Mennonites who arrived at the Molochna River remembered seeing kurgan burial mounds, Scythian graves dating from the seventh- to third-century BC, protruding from the flat steppe.[67] As Warkentin and Krehbiel would be reminded seventy years later by heaps of bones on the prairie, the grasslands were not so much vacant as haunted by successive removals and settlements. Theirs was only the most recent migration. Turkey wheat adapted well to the prairie, with its cold, arid climate and stiff soil cover so like that of the steppe. But there were other explanations for the Mennonites' success in the midst of a plague of locusts.

5 : SPACIOUS SKIES AND ECONOMIES OF SCALE

In the summer of 1874, Jacob Wiebe purchased twelve sections of railroad land fifteen miles northwest of Peabody, Kansas, and then waited for his family to arrive from their temporary lodging in Elkhart, Indiana. In the windy heat of summer, Wiebe's hopes sank. Grasshoppers devoured the few nearby wheat fields. Dust storms blew dried prairie grass through the village. But the land was bought, and so Wiebe resolved himself to the future. When his family arrived in mid-August, they loaded their ox carts and wagons and drove for hours through the dry tall grass of treeless prairie. When they reached a stake Wiebe had placed in the ground, he stopped the wagons. "Why are we stopping here?" his wife asked. "We're going to live here," he replied. Then she began to cry.[1]

Her tears were justified: within three years Wiebe and his wife returned to Crimea, their carefully planned village of Gnadenau declared a failure. Shortly afterward, new settlers had destroyed the houses lining the narrow central street and plowed up the narrow strips of farmland situated behind them. In their place were broad tracts of farmland settled by individual landholders. Communal living had failed.

Yet the new landholders were also Mennonites, and on the whole their experiment in the Midwest was a celebrated success. American journalists flocked to Mennonite farms in Kansas, fascinated by their ways of life. Within a decade the Mennonite settlements in Kansas appeared an "oasis in the desert," as one Russian writer described their counterparts on the southern steppe.[2] Their prosperity was especially noteworthy given the widespread poverty and indebtedness of neighboring homesteaders. The visiting journalists asked, as many have since: what allowed the Mennonites to succeed where so many settlers failed?

One rendition of the Mennonite story recapitulates the cherished myth of the American dream: a providential faith in the skill, hard work, and success of willing immigrants. The story of Turkey Red Wheat distills that heritage into a single artifact: the seed that made the United States a breadbasket to the world.

In the stories told about the triumph of agriculture in the plains, the seed itself has become a proxy for unspecified factors: displacement, labor and social organization, agronomy, state grants and assistance, trade networks, and land quality.

Postbellum Kansas was made safe for capitalized homesteading in ways quite similar to the southwestern steppe annexed by Catherine II in the context of the Russo-Turkish Wars. A border territory characterized by conflicts so violent they earned the moniker "Bleeding Kansas" in the late 1850s, Kansas had seen the end of constitutional skirmishes over the legality of slavery by the time the Mennonites arrived in the wake of the Civil War. Fleeing the Russian state as they had fled the Prussian a century before, the Mennonites were colonizers, the shock troops of a free labor economy, agents of imperial expansion on the North American continent as they had been in Russia. Settlers of means received grants and privileges from the state or railroad corporations favored by it. They settled among Plains Indians as they had among Tatar pastoralists. This pattern of homesteading was something other than the paradise of yeomanry Thomas Jefferson imagined: a patchwork of widowed landscapes, emptied reservations, and lands monopolized by railroad corporations. Mennonites combined cooperative social organization with a capitalistic orientation, departing from the serfdom of the steppe, the plantation slavery of the American South, and the land tenancy arrangements that emerged in the wake of both.

Wealth and large-scale landownership were the primary factors allowing for profitable cultivation of cereals in the Midwest. The Mennonites were rich and well organized compared to the many homesteaders who traveled west in search of land and livelihood. The favorable terms of the Kansas Mennonites' contracts with the Santa Fe Railroad Company enabled them to farm sizeable plots of good land. Unlike their neighbors, they did not remain tethered to creditors, forfeiting the season's profits to pay for the seeds and equipment required to grow their crops. What allowed Mennonites to be successful cultivators was only in part the hard red winter wheat they acquired, even as its acquisition obscured the long history of improvement embodied in the seed itself. For seeds are not merely agricultural inputs, but symbols of prosperity and bounty masking the political-economic requirements of cultivation.

By the late 1870s, the plight of Kansas homesteaders was sufficiently extreme to attract national attention. A journalist for the *Atlantic Monthly* arrived at harvest. He found the greater proportion of farmers beaten by indebtedness,

Figure 5.1. C. R. Voth and crew threshing in Marion County, Kansas, 1880–1920. Kansas State Historical Society.

having taken out credit to buy land and failed to recoup the cost with its products. Many homesteaders were too poor to afford a wagon or a cart.[3]

In the towns, he found unemployed Negro laborers who attributed the lack of work to the mechanization of agricultural labor. Henry King, writing on the picturesque features of Kansas farming in the same moment, celebrated the efficiency of the reaper, self-raker, and self-binder, the self-binder a "ghostly marvel of a thing, with the single sinister arm, tossing the finished sheaves from it in such a nervous, spiteful, feminine style." He added: "I wonder what Solomon would have thought of the self-binder?"[4] Christian Krehbiel, with spiritual credentials approaching Solomon's, weighed in his old age: "Although I enjoyed the hard work," he reflected, "I am very glad that machines have taken the place of the scythe and the cradle."[5] Yet for most farmers, crippled with debt, access to such marvels was a fantasy. Again and again the journalist encountered this "tale of wretchedness" among farmers heading farther west across the unbroken prairie to the Washington Territory.

If he witnessed misery and poverty, however, he also observed a diversity of settlements and considerable disparities in wealth. The towns in the Arkansas

River Valley near the railroads fared the best, especially the Mennonite settlements west of Topeka. Everything about the Mennonite settlements stood in opposition to the desperation surrounding them. Groves of fruit and shade trees surrounded the farm buildings, which elsewhere stood unsheltered and unadorned. Many farmers attempted mulberry hedges for silkworms. Vegetables, grapes, and flowers flourished. The fields were well tilled and cared for, producing the best crops of wheat and barley yet seen.[6]

To the journalist from the *Atlantic Monthly*, the Kansas Mennonites had proved success on the plains was possible. They had "shown how comfortable homes may be created in a short time by intelligent industry," he wrote, "assisted by capital sufficient to make a good start with buildings, tools, and seed upon a small piece of ground." This capital enabled the settler "to live two or more years without returns from the land cultivated." He added that "good and intelligent cultivation" was essential to managing climate and insects, which appeared insuperable to less experienced farmers.

In addition to start-up capital and industry, the similarity of the Russian steppe to the midwestern prairie aided Mennonite cultivators. "In Russia," he observed, "they were all farmers, and in coming to this country they have brought with them their life's experience in agriculture, under conditions of climate and soil not altogether unlike those of Kansas; and also many of the tools there used, though they are adopting our improved implements of husbandry." According to this journalist, the basic formula for success was start-up capital for two years plus intelligent, experienced cultivation and a community ethic.

Other journalists traveling the Midwest found similar patterns of settlement patterns and prosperity. One writer, E. V. Smalley, who traveled through the northern prairie, noted that the Mennonites were the only settlers to group houses into villages several miles apart.[7] Another followed the path of early immigrants up the Red River Valley to Manitoba and found large tracts of farmland settled by Dutch farmers on land sold to them by the Northern Pacific Railroad. Divisions of 5,000 acres were subdivided again to 2,500, operated by superintendent, foreman, and gang labor. Farther north, near Manitoba, just west of the old Indian settlement of Pembina, on the British side, Mennonite homesteaders threshed grain by horse hooves while girls gathered and winnowed it. They farmed smaller but still sizeable plots of land. The journalist remarked that Mennonites were almost without exception well to do.[8]

Yet it was not so for the Wiebes, whose village of Gnadenau failed within two years. What went wrong?

When the Molochna immigrants arrived in the fall of 1874, Kansas was set abuzz with talk of their sartorial choices, eating habits, and avid trade in provisions and farm implements. Early in September, large crowds gathered to see six hundred arriving Mennonites dressed in simple homespun garments of coarse wool. A reporter from the *Topeka Commonwealth* documented the scene with some relish. He observed that the women and children wore "funny old handkerchiefs tied around their heads" and had brought with them huge tin pans, crockery, and baskets "soon groaning under loads of bread, cheese, and sausage." The bread they consumed in huge pieces "with a rapidity which augured well for their digestion." He was no more impressed with the fashion of the men, who evidently possessed "conscientious scruples against wearing clothes that fit them, the idea appearing to be to get all the cloth you can for the money." Their vests extended nearly to their knees, and their trousers possessed "an alarming amount of slack." Strangest of all were the flat cloth caps they pulled off to salute any person, a custom the writer thought improbable to survive in Kansas, where, as the saying went, "nobody respects nothing."[9]

Even the poorest of Mennonites came with some advantages, buffeted by the gift tickets and temporary facilities provided by the railroads. Of the roughly 10,000 Mennonites who had settled west of Topeka, the *Atlantic Monthly* noted that "all have come with some means, the poorest of them having an average, according to the best information obtainable, of at least $1,500 each, while others have brought as much as $100,000." This estimate was inflated and failed to account for costs of migration. A Russian writer interviewing Mennonite immigrants in 1878 provided a more fine-grained assessment of their wealth, estimating the cost of emigration at around $700. The settlers he interviewed paid one ruble and seventy-five kopeks per dollar, with the total resettlement expenses approximately one hundred rubles per person. Assuming a family consisted of five persons, each family took 1,725 rubles from Russia, not counting the value of the movable property, which many transported almost in its entirety. The settlers he interviewed bought land at $2 per acre on credit to be repaid within eleven years. At the time of his visit, nine-tenths of all the settlers had paid all their debts.[10]

Nevertheless, many settlers came without cash reserves, with many having been forced to sell their land at a significant loss. Still others in the Polish section of Prussia had been settled on state lands, owning nothing but the profit from their crops. They came with no funds. Wiebe's Crimean community was quite poor as well, with some even owing their passage money. Christian Krehbiel persuaded Bernard Warkentin to lend Wiebe one thousand dollars, and Cornelius Jansen and Jacob Funk lent him another thousand each.[11]

These loans were evidence of the community networks Mennonites forged to support other congregations and aid impoverished immigrants. Some of the oldest Mennonite congregations from Lancaster, Pennsylvania, provided relief funds through the Board of Guardians, directed in part by Bernard Warkentin and Christian Krehbiel. When a hundred families from Volhynia, in the Polish section of Russia, arrived without warning or provisions in the fall of 1875, Warkentin and others interceded. The Santa Fe Railroad Company had left the families in a storehouse in Florence, Kansas. Warkentin wrote to Goerz, overcome: "I don't know what to do—I have paid seventy dollars for bread since we left St. Louis and have offered $400–$500 to pay for ovens and the groceries needed immediately." The people of Florence had refused to receive the migrants, denying even the use of their buildings; so 120 families with small children were living in freight cars as winter set in.

Warkentin and Krehbiel arranged provisions and persuaded the eastern aid committees in Pennsylvania to provide support. Meanwhile, they lobbied the railroad company to provide forty acres each with no demand of payment for five years. They asked settlers with unoccupied lands to build housing and take families on as tenants and new settlers to build additional housing for needy families in exchange for their labor. Later, critics in other congregations charged the Board of Guardians with being too worldly, profiting from the misfortune of starving settlers.[12] The penniless migrants found shelter and food. But the critics were right that one effect of this bargain was to draw them into organized economic activity profiting their landlords and fellow congregants.

This organized economic activity tapped into new national markets yoking midwestern settlers to eastern financiers. Settlers who could afford them ordered ready-made houses and barns from a trading house in Chicago ranging from two hundred to seven hundred dollars in cost, depending on the size. These arrived disassembled by rail and were assembled by workers from the trading house. Settlers used the remainder of their funds to acquire workers, draft animals, dairy cattle, pigs, poultry, and machines. Larger machines such as seed drills for cereals and maize, reapers, and harrows were bought on credit.[13]

Wiebe's settlement looked quite different in comparison, far more similar to the model villages specified by Johann Cornies on the southern steppe. Wiebe built his house in the Russian style, with stables adjacent to the rooms and the granaries upstairs. Gnadenau as a whole consisted of nine sections of land of one square mile each, arranged as a perfect square, three miles on each side. He laid out the village according to the communal model of Mennonite villages, with a central road lined with houses on the north and farmland cut behind them in narrow strips. On the south side, good and bad land was equally distributed, also carved into strips. Streams and ponds further divided small plots.[14]

Wiebe clung to the principles and divisions of communal land tenure, while most Mennonite settlements in Kansas transitioned rapidly to private property holdings. His was not an incidental preference. The Wiebes belonged to the Kleine Gemeinde congregation, a conservative faction that opposed the increasing worldliness of many Mennonite settlements. In Kansas, he tried to recreate the perfect community imperiled by Russification policies and pressures on landholdings.

Gnadenau was a novelty for American observers, who wondered at the socialist experiment in their midst. "For some time, the entire American press was occupied by the Russian communists," wrote one Russian writer traveling through Kansas in 1878. "There appeared countless magazine and newspaper articles, as well as brochures about the Russian communities." The writer, I. Dementyev, made an exhaustive tour of all the Mennonite settlements in Kansas, including the failed community of Gnadenau, which had been rebuilt on private holdings and was settled with forty families by the time of his arrival. He regarded communism in the United States a brief experiment: "after weighing all the chances pro and contra, the practical Yankees unanimously decided that it was not suitable and unprofitable, and the matter was buried forever."[15]

Dementyev took a special interest in the wholesale failure of Russian communal land tenure, identifying a number of problems with the communal land system for doing business in the United States. Above all, there were no fallow fields. All land was cultivated annually. Since farmers cultivated crops with very different maturity cycles simultaneously, access to land strips for harvest and pasture was difficult. The larger machines, such as self-binding reapers, were too big to drive along the narrow strips. Neither could machines drive sheaves across the strips for threshing in the center of the village.

Prairie fires also made Wiebe's settlement pattern a risky one. The Wiebes' neighbors advised them against clustering houses in a row amid the tall prairie grass. They showed Wiebe how to plow furrows around the village as breaks against fire, and how to burn the grass between the furrows.[16]

On other farms in the area, labor-saving machinery facilitated the cultivation of large tracts of land. Light plows with two plowshares drawn by two horses or mules replaced the old, heavy Russian ones with their four pairs of oxen. Self-binding reapers replaced the sickle, "viewed by the Americans with amazement as something barbaric and horrible." At least according to Dementyev, the switch made for "an easy and happy job." Meanwhile, "beautiful and light cars on leaf springs" replaced "the heavy, creaky vans, quite a heavy load by themselves."

The relatively short distance to market and low price of commodities encouraged increased production. According to Dementyev, the farthest distance to market in Kansas was twenty miles, rather than the seventy-five along a terrible road to the port of Berdiansk. The size of land plots had increased accordingly. Whereas many of the first settlers, accustomed to manual labor, had purchased plots of eighty acres, by the time of Dementyev's visit, the average plot size was 320 or more acres. According to one of Dementyev's subjects, Pastor Buller, local yields were at least twice as large as in Russia. The main product of Mennonite farms was winter wheat, along with small quantities of spring wheat and maize, the latter for pig feed. Farmers cultivated sugarcane, sorghum, broomcorn, and vegetables for local consumption.

Among the farms Dementyev visited was Jacob Funk's, one of the first group of settlers who traveled with Krehbiel and Warkentin. His farm was located four miles from the train station in the Cottonwood River valley. Nearly everywhere Dementyev noted the presence of huge elevators for grain storage. He wrote admiringly of the large stone houses, extensive barns, orchards, and vineyards on the farm. Dementyev noted that Funk, as "a man of substance," had settled separately from his former countrymen. Fourteen poor families from the Volga who had arrived with no means farmed his land. Funk, a wealthy landowner in Crimea, had transferred his prosperity to Kansas by means of capital, landholding, and a ready supply of labor.

By all accounts, the Mennonites became students of agricultural innovation in Kansas. According to Pastor Buller, the Mennonites' prominence as agriculturalists in Russia had impeded improvements. There they were regarded as teachers and had only the subsistence farming of Russian peasants as a point of comparison. "Yet here," he claimed, "we happened to be the total ignoramuses and we need to learn permanently and keep a watchful eye for all the innovations. Otherwise we will continually lag behind our neighbors."

Buller was amazed with the rapidity of innovation, with a new reaper design "almost every time the harvest comes." The reaper the settlers found when they arrived in 1874 required gathering the mown crop with a rake. Shortly afterward, a new one replaced it, which bound the wheat into neat sheaves. The next model had a platform for the binders, reducing the number of required workers from five to two. In 1876, the self-binding reaper eliminated the need for manual binding altogether, such that only a driver and a pair of horses could do work formerly requiring many hands. "Hardly have we stomached this news," Buller wondered, before the introduction of the header, which cut only the ears and deposited them in a wagon traveling beside it, eliminating the need for sheaf handling. According to Buller, changes happened every year, and only by keeping a watchful eye and attending "competitions and exhibitions to learn from American neighbor farmers" could Mennonites keep pace. Meanwhile, Buller noted, the former isolation of Mennonite communities proved practically impossible to maintain.

The abandonment of communal land management was an ideological shift as well as a practical one. Dementyev wondered that Yankee ambition had powerfully altered German habits in a single generation, whereas the same people had failed to acquire a word of Russian in a whole century. "It is impossible to find a better proof of our utter inability to Russify other nationalities," he offered. "Some . . . have moved to Russia as far back as in the times of Catherine the Great and lived among the Russian people for over a hundred years, yet remained Germans from head to toes, except for the use of grey shchi [a cabbage soup] and Russian stoves." Meanwhile, after only four years in America, Dementyev wondered, "the restless Yankees have already infected with agility and entrepreneurship even these imperturbable sons of Teutonia." He also noted their readiness to abandon their favored farming practices for American ones. "The spirit of entrepreneurship has woke up in them, and many, having shaken off the century of stupor, have already become engaged in trade and industry." If Dementyev underestimated the entrepreneurship of Mennonites in the southern steppe, he nevertheless registered a significant ideological transformation associated with large-scale cultivation on the prairie.

North of the border, the communal system held on longer. Mennonite settlements near Winnipeg in the Eastern Reserve of Manitoba adopted the traditional model, grouping themselves into villages of fifteen to twenty families. Steinbach, consisting of eighteen families, was a typical example, continuing as a prominent trading center past the Eastern Reserve's abolition of the farm village system in 1909. Houses lined a central street, ninety-nine feet wide and one mile long, surrounded by ten-acre lots for gardens and barnyards. Farms were narrow strips a half mile to a mile in length. Farmers in these villages rarely owned the strips they cultivated, but rather a regular plot of 160 acres in the village as a whole. Neighbors thus farmed each other's land, bound by their own regulations rather than legal title. Hayfields and pasture, too, were held in common.[17]

Yet ultimately in Canada, as in the United States, large-scale wheat cultivation required abandoning the village pattern. The late advent of the railroad to Manitoba delayed the dissolution of the village system. By the mid-1880s, only several years after the railroad arrived, village life disintegrated quickly. Property boundaries of sectional surveys, rather than strips laid out for cultivation by community consent, provided the legal basis for the management of property, including the mortgaging of homesteads required to acquire equipment for large-scale grain cultivation. If an indebted settler lost his 160 acres, and those acres were divided among several strips of land farmed by his neighbors, the village as a whole was threatened.[18]

In a world without railroads and extensive farm implements, the village system was efficient, allowing for wide communication without interrupting farming operations, herding without barbed wire, and production of straw and manure for burning stoves. In the seven years before the railroad arrived in Manitoba, Mennonite settlers in the village system produced more, and more efficiently, than their neighbors.[19]

But this was not the world in which Kansas Mennonites found themselves.

By the time the journalist from the *Atlantic Monthly* visited in 1879, the Wiebes had returned to Russia, and the village as they constructed it had been demolished, as had nearly all the others. The conservative Alexanderwohl congregation from the Molochna River area, too, had tried the village system, with nine districts of strip farms built along a road of one mile. These also lasted only a few years, with all but Hochfeld broken up and replaced by American-style farms with a single farmhouse.[20] There were simply too many arguments against the communal system. "Indeed," Dementyev concluded, "these reasons were so numerous that the poor defenders of the communal principle did not know what to do, so the village had to be disassembled and the building done on individual sites."[21]

Even in its traditional form, however, Mennonite property organization was cooperative but not communal. This flexibility allowed farmers to scale up production for national and international markets. Smaller and more isolated Hutterite and Amish communities provide a point of contrast. These communities often retained communal social and economic systems. Neither participated in large-scale production. The Hutterite Bruderhof held all property in common among its members, and it succeeded partly on the basis of underproduction. Periods of prosperity amplified inequality and generated conflicts within the Bruderhof, often resulting in fission into multiple new communities. Hutterites responded to population pressures, resource scarcity, and technological innovations, but never with an intention of producing surplus.[22] The Amish, who unlike the Hutterites rejected all technological advances, adopted similarly low-input agricultural methods, reliant on animal and hand labor in lieu of machinery or fertilizers. Their social organization determined their technological choices in addition to their priorities for production, leading communities to retain a diversified farming economy for household and regional markets in lieu of monocultural production associated with values of entrepreneurship and business.[23]

In contrast, observers considered the Mennonite cooperative spirit a boon to increased production. "There are strong bonds of sympathy between them," the *Atlantic Monthly* journalist reported, "and they are helpful to one another." If the loans and shelter extended to penniless settlers saved them from ruin, they also guaranteed ready labor for more affluent Mennonite settlers. Although communal land management failed, cooperative institutions flourished, including charities, hospitals, orphan homes, and insurance companies. Bernard Warkentin, a prominent mill owner and grain elevator operator, led several such institutions, including Bethel College, the Bethel Deaconess Hospital, and the Mennonite Mutual Fire Insurance Company. While elsewhere farmers decried the abuses of grain elevator operators in partnering with the railroad and giving farmers a bad rate for their crops, Mennonite wheat culture was vertically integrated in ways that controlled product quality and mitigated labor unrest. And it embraced the politics of scale.[24]

For many Mennonites, there was no contradiction between capitalist and cooperative values. Warkentin was one of these, as had been Johann Cornies. On the basis of his early immigration, connections, and wealth, he became a successful businessman. After his voyage across the west as a scout in 1871, he migrated to Halstead with Krehbiel's Summerfield community in 1874. He built Halstead's first gristmill on the Little Arkansas River. With childhood friend David Goerz, he began publishing the German-language magazine *Zur Heimath* in 1875. The same year, he married the daughter of a successful mill owner and grain elevator operator. In 1886, following travel to southern Russia, Warkentin founded the Newton Milling and Elevator Company, expanding to Oklahoma in 1900 with the establishment of the Blackwell Milling and Elevator Company. By 1900, Warkentin had also established the Halstead State Bank and the Kansas State Bank of Newton, of which he was director and president. In short, he was a community leader and a person of power.[25] His business had expanded from a single mill to a major interregional business for national and international markets.

Mark Carleton's story of Turkey wheat's origins looks somewhat different in light of the foregoing history. When Carleton attempted to give credit where credit was due, he included the names of the people with whom he had close dealings: Warkentin, Krehbiel, and Schmidt, a successful miller, real estate agent, and railroad agent. All three had worked together closely in arranging the purchase of the land from the Santa Fe Railroad for Mennonite settlement.

Although Carleton credited Mennonites in general, and Warkentin in particular, with the introduction of Turkey wheat, it was less a question of conveyance than cumulative labor. The name "Turkey wheat" was already in use in Kansas by the time Warkentin immigrated, but the efforts of Warkentin and others made it a primary cultivated variety. The wheat required changes in milling technology to accommodate its toughness. Warkentin both led and profited from these developments. His commitment to securing a quality product for his mills meant acquiring new seed from Russia whenever possible. His efforts included several trips to Russia, imports of seed grain in the 1880s and 1890s, trials with different wheat varieties near Halstead, and ultimately a partnership with the USDA's Mark Carleton to import new varieties.

In succeeding years, USDA agronomists would favor Mennonites in trials of new seeds. When Mark Carleton went to Russia in 1900, he told Warkentin seeds would be shipped mainly to agricultural experiment stations, but also to "responsible and intelligent farmers" such as could be identified by the USDA. Here Carleton alluded to techniques of cultivation such as those practiced by the Mennonites.

As Carleton suggested, agronomy was at least as important as seeds in determining the success of a crop, a lesson that has as much relevance for contemporary development professionals as it does for historians of agriculture and biotechnology. The Mennonites used dry farming techniques, including deep plowing, which was tried and proven on the Russian steppe. They enriched the soil with nitrogen-fixing cover crops such as clover and timothy and paid special attention to manure. If they regarded the land as a renewable resource rather than an exhaustible one, they nevertheless contributed to the transformation of the wild grass prairie into a grain depot: an undertaking with major environmental consequences.

The locusts were a special trial. Mennonites had experience with them in Russia. In addition to cutting ditches, farmers swept them manually to the prairie margin, which was set on fire. They also set fire barriers on the fields to stop migrating insects. They did not participate in the so-called grasshopper conventions of 1877, instead keeping to themselves and their learned techniques for battling pests.[26]

But wealth rather than agronomy was the primary predictor of prosperity in blight. Usually the pests came in August, when the wheat was already harvested. In 1876, they came in May and wiped out all the wheat. Mennonites of means replanted spring wheat and still managed a good crop.[27] Most did not have this luxury. In grasshopper years, these farmers lost their crops, while in other years they managed to repay their debts. Those who had mortgaged their lands, however, lost them.[28] The kind of responsibility and intelligence Carleton specified also required resources.

Private property and state grants and privileges provided the engine of agricultural development in the United States. Myths of national bounty, heritage, and innovation mask these political economic arrangements, in part by refiguring the seed as an object of natural advantage, cultural property, or research and development, rather than a product of labor shared across space and time. As Carleton sought to attribute credit to Mennonite introducers, he was also beginning to recast seeds as objects of innovation.

By the time Jacob Allen Clark and Carleton Roy Ball assembled their encyclopedic classification of American wheat varieties in 1922, Turkey Red had spawned many other hard red winter and spring wheats through deliberate crossing.[29] The USDA's Office of Dry-Land Agriculture in the Bureau of Plant Industry, established in 1905, used single-line selection and hybridization to develop new varieties from Carleton's imports, including P-762 and Kanred.[30]

Clark and Ball articulated the need for a comprehensive classification of wheat varieties at the outset of their volume; given the many varieties now in development and on the market, agronomists, breeders, and those in the grain trade required order for continued improvement. According to Clark and Ball, they also needed to protect themselves against threats of duplication and fraud, new threats that indicated commerce was displacing traditional seed-sharing arrangements as a mode of biological production. Clark and Ball included a summary of foreign and domestic investigations to date, from the ancient histories of Pliny to Linnaean and Lamarckian taxonomics to the recent efforts of American improvers such as the Ohio secretary of agriculture John Hancock Klippart, Sereno Edwards Todd, Joseph Buckner Killebrew, and the USDA's own Mark Carleton.

While they limited their study to the principal cultivated varieties and focused especially on nurseries and experiment stations, they nevertheless struggled to consolidate the numerous varieties under cultivation. Even John Hancock Klippart's 1859 synthesis *The Wheat Plant* had included some ninety varieties in wide use, which he divided into Tinged, Whitish and Reddish Beardless, and Bearded subsets. These went by both commercial and folk names: for example, Duke William, Eclipse, Le Couteur's compact, and Chinese spring wheat. Notably both "Turkey large red" and "Caucasian red" were listed, suggesting that perhaps small amounts of hard red winter wheat from the Black Sea and Caucasus regions had been tried in the middle states as early as the 1850s.[31]

Clark and Ball developed a lexicon for breeders, pegging plant morphology to varietal nomenclature and the emerging property rights it implied. Yet the linkage belied the extent to which the adopted names were a palimpsest of improvisational, customary, and commercial undertakings to steward and improve seed. In an attempt to simplify nomenclature, Clark and Ball reduced descriptive appellations to single words: "Turkey Red" became "Turkey." Each variety had a paragraph of history provided, which was anecdotal in the extreme, peppered with the words "reportedly," "reputedly," and "allegedly." If the *who*, *where*, and *when* were undetermined, this was because never before had seeds been subject to the logic of title, fixity, and property, assuming a generational rate of change over time. Even Carleton's identification of Warkentin as the introducer of Turkey wheat was a belated one, a sign that biological innovation was becoming a question of individual advance, not collaborative labor.

In this climate of standardization and institutionalized research, international collaboration between experiment stations flourished, including subsequent partnerships between Soviets and Americans. A year after Carleton toured southern Russia, Gifford Pinchot of the United States Forest Service admired the tree breaks in the southern steppe. Pinchot's observations preceded midwestern American adoption of the method by some twenty years. The environmental crisis of the steppe in the 1890s presaged the American dust bowl of the 1930s, and Americans looked to Russians for lessons on how to address the consequences of transforming grasslands for pasture and arable husbandry.[32]

Carleton and others sought well-adapted seed varieties from abroad, a deliberate exercise in internationalism that masked the extent to which the prairie was already cosmopolitan. Meanwhile, the Japanese imported many of the same varieties from the steppes and the plains, including Turkey wheat. One descendent was Norin 10, the semidwarf variety Cecil Salmon collected from occupied Japan in 1945. Salmon was a professor of farm crops at the Kansas Agricultural Experiment Station, where Carleton had worked, and an authority on wheat cultivation in the United States. Ultimately, Norman Borlaug crossed Norin 10 with Mexican varieties, enabling world cultivation on a scale theretofore unimaginable.

But the previous century's settlers had experienced unprecedented scale as well.

The force of nature stunned all the travelers on the western plains as they gazed through the windows of coaches and trains. When Krehbiel and Warkentin first saw a storm, they did not recognize it. "What is that?" asked the group when all at once "a dark mass arose in the west" with seagulls driven before it. The rest of the trip proceeded in a cold wind.

Those trails, and the prairie itself, were gone by the time Katharine Lee Bates crossed the plains. She saw nothing but amber waves of grain: fields of hard red winter wheat, straight from the Crimea. She was taken with the sheer bigness of it all, which she equated with national greatness. Henry King, a city slicker and national booster who thought Kansas farming nothing if not picturesque, was inclined to agree: "The farms are large, you will observe, and growing larger," he wrote for *Scribner's*, "as if they had caught something of the nature of those infinite skies."[33] The spacious skies called forth economies of scale.

Others were not so sure. Henry Van Dyke, the journalist who went north along the Red River for *Harper's*, traveled with an illustrator who tried and failed to capture the landscape. Van Dyke wondered whether large farming served the country well. "It absorbs great tracts of land, and keeps out smaller farmers. It employs tramps, who vanish when the harvest is over, instead of increasing the permanent population. It exhausts the land. The cultivation is very shallow. There is no rotation of crops. Everything is taken from the ground; nothing is returned to it. Even the straw is burned. The result of this is that the average crop from any given acre grows smaller every year, and it is simply a question of time under the present system how long it will take to exhaust the land."[34]

Van Dyke, with many others, anticipated the many critics of the so-called industrial agriculture of the twentieth century. "As we looked out over the great plain, and slowly took in the extent, the fertility, the ease of cultivation," he wrote, "we echoed the local brag: 'This is a big country, and don't you forget it.'"

"Yes," said the illustrator, "that is the trouble; it's too big. I can't get it on canvas. A man might as well try to paint a dead calm in mid-ocean."

Field Notes

"INDIGENOUS KNOWLEDGE"

DIVERSITY AND ENDANGERMENT

Mennonites brought capital, grain, and Russian thistle to the American Midwest, and they encountered prosperity, grasshoppers, or drought in the field. Migrants flourish and falter based on fortunes: those they bring with them, and those they encounter in their travels, for good or ill. When Mark Carleton tried to account for the success of Mennonite cultivators, he prioritized not the capital but the seed they brought with them, spinning a myth of origins for large-scale grain production that prioritized inputs over agronomy. Moreover, in setting out to follow Warkentin's path to the "original seed," Carleton recast a history of migration as one of tradition, and a history of movement as one of stasis. If his logic drew support from several centuries of Euro-American collection and transplantation of plant and animal resources for economic gain, it also anticipated the next century's shift from species to genes as objects of conservation and utilization.[1] Both traditions regarded plants as resources for improvement and sketched a geography of global wealth in which sources of diversity were static.

When Nikolai Vavilov cited Turkey Red Wheat as an instance of ancient Crimean seed stocks contributing to modern American agriculture, he was in the process of formulating his influential theories on biodiversity and the origins of cultivated plants. Based on his travels, he theorized a geographic distribution of genes and identified "centers of origin" for cultivated plants based on high levels of diversity from the regions in which he collected.[2]

Vavilov also proposed a positive relationship between biological and linguistic diversity. The highlands of Central Asia, and especially the Panj River valleys of the Pamir Mountains, were a center of Vavilov's exploration in the 1920s and instrumental to his theories about the origins of cultivated plants. During his travels, he found that naming practices for food plants varied between highland and lowland inhabitants of the Pamirs, positing that variation in dialect and linguistic marking indicated localized preferences for selection of desirable traits in cultivated plants. In his analysis, individualized language suggested geographic isolation, which in turn suggested the presence of plants with the same characteristics.[3]

Vavilov's theories of biocultural diversity set plant genetic resource conservation on a path of privileging indigeneity as the locus of biodiversity. This reverence for stasis and isolation generated an ironic if not paradoxical quest: to reach areas untouched by development in order to preserve that which development would imperil. In Vavilov's research for the Soviet state, as in Carleton's for the United States, collected seeds were useful in large part because they enabled the production of varieties for large-scale cultivation. The project of "genetic modernization" in which Vavilov and other Soviet scientists were engaged implied reforms of peasant agriculture, and arguably of peasants themselves.[4]

In the twenty-first century, modern varieties of seeds reach remote regions, if not by markets, then by aid. For even as it allows for the possibility of preservation and tradition so valued by collectors, stasis is no defense against poverty, or threats of violence and deprivation. These conditions raise complex questions about how preservation projects engage with underdeveloped regions identified as centers of biodiversity. Although the team of plant genetic resource specialists I accompanied targeted watersheds and pockets of biocultural diversity, expedition routes also tracked histories of instability and conflict in the republics of the former Soviet Union.

The Pamir Mountains remain a focus of plant genetic resource collection. Pamiri people speak indigenous languages of the Badakhshan region spanning Tajikistan and Afghanistan, generally classified as members of the eastern Iranian language group. These languages have a number of variants and few remaining speakers.[5] According to the Food and Agriculture Organization's guidelines, Pamiri people qualify as both indigenous and traditional. Yet their livelihoods are far from secure.

Tajikistan, bordered by Afghanistan, Uzbekistan, Kyrgyzstan, and China, is the poorest of the fifteen post-Soviet republics, with the lowest GDP, lack of employment opportunities, and industrial and agricultural production crippled by the civil war of the 1990s. Signs warning of land mines routinely interrupted our collections. Yet it was the very poverty and remoteness of these regions that made them preserves of the traditional varieties of seeds pursued by collectors.

Tajikistan remains dependent on international aid agencies for basic subsistence, and the most active of these agencies is the Aga Khan Foundation, directed by the Aga Khan IV, an international business magnate who is also the imam of the Ismaili Shiites, the majority population in the Pamirs. Agricul-

Figure FN3.1. Soviet tank collapsed on embankment over wheat field on Panj River border, Tajikistan and Afghanistan. The Tajikistani civil war from 1991 to 1997 decimated agriculture in the area, which has only recently been cleared of land mines. Photo by Courtney Fullilove.

ture in the area was decimated by conflict. Much of the area has only recently been cleared of land mines, its fields dotted with warning signs and Russian tanks collapsed over embankments. One of the largest sectors of Tajikistan's economy may be the black market traffic in opium from Afghanistan to neighboring countries: unmarked trucks careen along the roads near the Panj River bordering Afghanistan.

For our team, this evidence of military and narcotic mobilization was also an indication that we would find few crops of interest. These heavily trafficked regions were near roadways linking farms to major cities. Seed stocks were likely to be a mishmash of Canadian, Australian, and Russian varieties provided by aid agencies in the wake of the conflict, mixed with traditional seed stocks via threshing on roadways postharvest. The result was a soup of degraded hybrids masquerading as landraces. Numerous villages nearby cultivated predominantly commercial varieties of wheat brought from the hub of Murghab, the only town in the region with an airport. The team moved on, up the river valleys and away from paved roads.

The team was sixty miles up the Bartang River valley in the Pamir Mountains of Tajikistan when we found the einkorn wheat in a tree. There was a small satchel of it stowed in a low-hanging branch, waiting for us. The villagers in Chadud had been advised of our visit, and a neighbor led us over the river's footbridge to the village, through the fields, past a mud grain silo, to the home of a farmer who was away on business. In a tree in his courtyard hung a small cluster of what appeared to be two-row barley, but on further examination was a diploid wheat, containing two chromosomes rather than the six of common (hexaploid) bread wheat. Diploid (einkorn) wheats grow cultivated (*Triticum monococcum* L. subsp. *monococcum*) and in the wild (*Triticum monococcum* L. ssp. *aegilopoides*). The wild version has a brittle rachis, shattering easily into segments. This one was cultivated, with a sturdy stem supporting the spikes.[6] Cultivated einkorn wheat is typically considered a relic crop, occasionally used for animal feed.

The farmer who took us to the tree said the wheat had been brought from a smaller village farther up the valley, accessible only by foot, even more remote from commercial networks than the place to which we had walked. By all appearances, it was a find. But there was too much we didn't know about it. Perhaps the farmer in the nearby village had saved it because it was special, or displayed it as an ornament because it was useless, small in quantity and requiring hulling rather than threshing for consumption.

Whatever its utility for modern agriculture, however, this spike of wheat highlighted two of the most important events in the domestication of the wheat plant: the acquisition of a nonbrittle rachis, which this plant had, and free-threshing quality, which it did not. Whatever the questions of provenance, there was no question that the material was rare, and that its preservation was attributable to the relative isolation of the villages up the valley.

In retrospect, having so little information about the plant seems pitiable, but we worked through interpreters, and there was much else to collect. And arguably, knowing the details of the seed's acquisition was relatively unimportant. Even for standard varieties, few of the many vernacular names for the plant made their way into the database, as Pamiri was translated to Russian to English.

However much effort was made to document the provenance of the seed, ultimately cultural diversity took a backseat to genetic profiling in the process of collection: an ironic artifact of Vavilov's transition to the genome as a locus of preservation and a model of diversity. There is ongoing dispute among agronomists, breeders, and genetic resource curators about the extent to which provenance matters, and implicitly, whether the preservation of biodiversity requires any historical consciousness.

These considerations veil a grittier debate over credit and benefits sharing

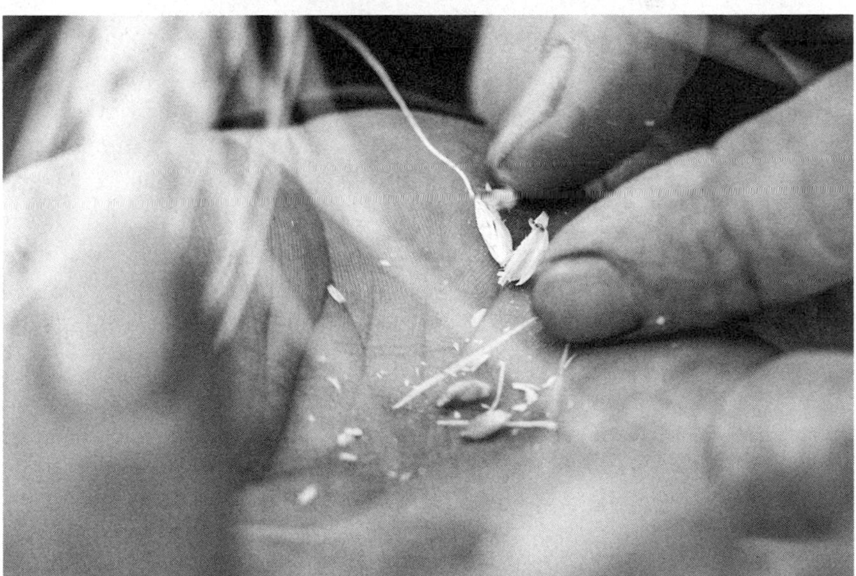

Figure FN3.2. Diploid wheat (*A*, *B*), silo (*C*), and mill (*D*) in Chadud village, Bartang River valley, Badakhshan, Tajikistan. Photos by Courtney Fullilove.

in the development of improved varieties that emerged in response to the shortcomings of international plant genetic resource policies devised in the context of the Green Revolution. Who has the authority to give? How are improvers compensated? If one goal of our travels was the production of new commercial varieties, was the preservation of seed a contribution to diversity, or an erosion of it?

These tendentious political and ethical questions came to a head around antiglobalization movements in the 1990s, with critics charging that efforts to exploit global biodiversity amounted to neoimperialism, favoring commercial monocultures and capitalized breeding rather than local populations. As new laboratory techniques escalated the screening of plants for active chemical compounds usable in medicines, cultivars, and industrial chemicals, biologists and policy makers urging sustainable development coined the term "biodiversity prospecting" or "bioprospecting" to describe these efforts.[7] Beginning in the early 1980s, the Rural Advancement Foundation International (RAFI), an NGO advocating rural livelihoods, targeted the biotech industry's deployment of patents on organisms that remained public goods in their countries of origin.

In RAFI's analysis, agricultural and pharmaceutical companies in the Global North were using the cover of biodiversity preservation to amass for profit the plants that formed the basis of diets and medical traditions for peoples around the world. According to critics, multilateral environmental and trade

agreements only intensified these movements. While the 1992 Convention on Biodiversity (CBD)[8] asserted varied forms of traditional and indigenous knowledge, its framing in terms of stakeholders nevertheless elevated the interests of those with the most capital: states, biotechnology companies, and large NGOs. The formulation of Trade Related Intellectual Property Rights (TRIPS) in the Uruguay round of the Global Agreement on Trade and Tariffs (GATT) in 1998 made the supremacy of industry apparent.[9]

In one internationally publicized case, RAFI joined the Council of Indigenous Traditional Midwives and Healers of Chiapas, Mexico, in protesting a project on "Drug Discovery and Biodiversity among the Maya" funded by the International Cooperative Biodiversity Groups (ICBG) and directed by University of Georgia ethnobotanists Brent and Eloise Berlin. The ICBG is a consortium of federal agencies including the National Institutes of Health (NIH), the US Department of Agriculture (USDA), the National Science Foundation, and the United States Agency for International Development (USAID). In the years immediately after its inception in 1993, the ICBG supported projects in Mexico, Peru, Chile, Argentina, Panama, Suriname, Madagascar, Vietnam, Laos, Nigeria, Cameroon, and Costa Rica. According to RAFI's opposition, the ICBG's commercial partners included the transnational pharmaceutical and agrochemical companies Glaxo-Wellcome, Bristol Myers Squibb, Shaman Pharmaceuticals, Dow Elanco Agrosciences, Wyeth-Ayerst, American Cyanamid, and Monsanto: evidence that industrial imperatives of progress and profit reigned supreme over the preservation of biodiversity.

While RAFI and the council of midwives and healers vehemently opposed the research as theft of indigenous knowledge and resources, advocates regarded collaborative projects with benefits-sharing mechanisms regulated by the Convention on Biodiversity as legitimate projects to preserve and document cultural and biological diversity. They charged that the council of midwives and healers was in fact a trade group attempting to protect its market position, and that it had no license to speak for indigenous people, who might possess overlapping bodies of plant medical knowledge. Ultimately, the project folded in the face of fierce criticism in 2001, two years into its five-year funding cycle.[10]

ICBG Maya was not an isolated case, though its basis in the Convention on Biodiversity made it especially noteworthy as an object of attack. In another alleged case of corporate bioprospecting, activists argued that Eli Lilly owed citizens of Madagascar profits for the use of periwinkle to produce Vincristine, an anticancer drug. Partly in response to this case, environmental activists from the Rainforest Alliance in New York funded projects to shelter plant-based drugs from threats of development. Politicians too might invoke preservationist arguments, especially ones that served economic ends.[11] President George H. W. Bush signed the Pacific Yew Act into law to protect old growth stands in the Northwest for the production of Taxol, a Bristol Meyers Squibb patented cancer drug identified and developed under the auspices of the National Cancer Institute, the NIH, and the USDA.[12]

Implicit in these debates were uncertainties about the relationship between biological and cultural diversity, and of both to commerce. Policy makers regard plants as endangered and make rearguard attempts at their preservation, but how do they regard the people who subsist among them? Is the circulation of knowledge desirable, or does it compromise diversity? Does development necessarily erode diversity by discouraging isolation? Do international preservation initiatives oppose development or encourage it? Is it sufficient for a specimen to have "passport data" if it is collected from people who have none?

Imperfectly addressed by new benefits-sharing provisions in the CBD, the specter of Chiapas nevertheless haunts twenty-first-century biodiversity collecting enterprises, suggesting that preservation has always been coupled with exploitation. The proper relation between smallholder agriculture and commercial monoculture remains unresolved, as do broader questions of how local communities should interface with international markets. Linkages between rare language and rare plant nevertheless have generated new arguments for preservation, buttressed by environmental threats and the rhetoric of endangerment.

The concept of endangerment is not novel to the late twentieth century, however. In the midwestern American plains, making way for the plowman entailed the removal of bison, indigenous settlers, and diverse prairie grasses, some of which formed the basis of American botanic medicines. While prairie grasses were regarded primarily as energy sources for herbivores, they were also the basis of indigenous pharmacy. As Pekka Hämäläinen has noted, the Comanche ceased to use some hundred plants and the knowledge associated with them as they shifted their economy from plant gathering to hunting.[13]

Boosters and critics of agricultural development acknowledged these sacrifices. While settlers plowed at their peril, a few articulated varied notions of preservation, from the enclosure of parklands to ethics of stewardship. Fewer questioned how autonomous professional knowledge taking shape in the sciences and social sciences supported the exploitation of nature. John Uri Lloyd, a Cincinnati pharmacist who manufactured botanic medicines, was one of these few, and he is the subject of the next two chapters of this book. If environmental legislation, utilitarian arguments for bioprospecting, and activism on behalf of the rights of farmers and indigenous people remain ambiguous on the questions of what is to be preserved and why, these uncertainties are legacies of the unresolved relationship between ideologies of progress and preservation in the preceding century.

PART 3 : **PRESERVATION**

INDIGENOUS PLANTS AND THE PRESERVATION OF BIOCULTURAL DIVERSITY

(Cincinnati, Ohio; Smithland, Kentucky; Earth Core, n.d.)

6 : ELK'S WEED ON THE PRAIRIE

In the weeds, things looked different. A year after Carleton's return from Russia, winter wheat flourished where Wiebe had planted his stake in the prairie twenty-five years before. The grain had choked out the prairie grasses and wildflowers Wiebe found, but they reasserted themselves within the crops. Among the heads of wheat, purple flowers bloomed, ignorant of the new economy imposed on them.

Late in the summer, after the flowers had dried and blackened, Bud Payton, a forty-two-year-old teamster from Paola, Kansas, walked along the crop margin, plucking the plants and putting them in a bag. Afterward he carted them to the post office and shipped them to Ohio pharmacist John Uri Lloyd, who had contracted Payton to secure a supply of the plant for the manufacture of an analgesic ointment he called Echafolta. When Lloyd issued a circular in the *Medical Outlook* seeking supplies, he advised readers that it was "commonly known as 'nigger head'" and as such was "possibly a common weed to the people of your neighborhood."[1]

A weed, the saying goes, is a plant out of place. By the late nineteenth century, a wide range of indigenous plants grew out of place, a nuisance to the farmers whose livelihood depended on grain cultivation. Medicinal plants such as Lloyd sought were lumped with other native and foreign grasses that had been conveyed intentionally and accidentally to the prairie. These included Russian thistle (*Salsola pestifer* A. Nels.) brought by the Mennonites, along with their more desirable plants, and *Elymus repens*, or quack grass, advocated by some agronomists for its quality as a pasture grass. Crop growers had little interest in such projects, either as they related to the taxonomy of grasses or the expansion of pasturage. Intruders into crops were weeds, whether they were European invaders or indigenous ones.[2] The conversion of prairie to farmland, along with private landownership, fencing, and increasing labor costs, imperiled the wild grass prairie. A political economy prioritizing the stewardship of indigenous plants was not in the making.

The plant Lloyd pursued was purple coneflower, or *Echinacea angustifolia*, a member of the daisy family endemic to the prairies of the Midwest, from Texas to Montana and Saskatchewan to eastern Oklahoma, western Iowa, and western Minnesota. Two other species, *purpurea* and *pallida*, flourished in the Southeast and central Midwest, from Georgia west to Louisiana, and north to Virginia, the Ohio Valley, Michigan, Illinois, and Iowa. The various names referred to the plant's morphology. German botanist Konrad Mönch devised the genus's Latin name in 1794, revising Carl Linnaeus's earlier identification of the plant as *Rudbeckia purpurea*. Mönch chose the Greek "echinos" (hedgehog) to describe the plant's spiny, rounded seed head. The subspecies name, "angustifolia," means "narrow-leaved." "Purple coneflower" described the color and shape of the head as it bloomed. [3]

But *Echinacea* had many other names as well, as Lloyd's reference to the racist vernacular indicates. "Nigger Head" was a derogatory reference to the plant's status as a crop weed. "Black Samson" is a more puzzling moniker, and one of the most common.[4] The biblical Samson owed his power to his hair—a divine gift he squandered by being careless in his lust for women. After his lover Delilah betrayed him by revealing the source of his strength, Philistines blinded, imprisoned, and ultimately killed Samson.[5] Like the more derogatory "Nigger Head," "Black Samson" was a racist moniker, figuring the dried flower head as African hair, but in the latter designation, its bearer was a tragic figure with pseudo-magical powers rather than a simple object of derision. According to the story, although venal and vanquished, Samson had strength.

The names applied to the plant by Plains Indians made references to the plant's uses rather than its morphology. The Omahas and Poncas called them "mika-hi" or comb plant, because they used the seed heads to comb their hair, or sometimes "inshtogahte-hi," referring to its use to wash the eyes ("inshta"). The Pawnee called it "ksapitahako" (hand, to whirl), because of a game children played with it, or "saparidu hahts" (mushroom medicine), because of the shape of the head. "Kansas Snakeroot" referred not only to the configuration of the roots, but perhaps also to a range of medicinal uses. The label snakeroot was applied to multiple taxa of plants used to treat snakebites and numerous other afflictions. Some called it Elk's Weed or Elk Root, reputedly because Plains Indians learned of its medicinal properties by watching elks graze on it when sick or injured.[6]

When Lloyd pursued supplies of *Echinacea* for his branded pain-relief ointment, he drew Native American pharmacological practices into a tradition of Euro-American botanic medicine. Melvin Gilmore, an ethnobotanist working in the early twentieth century, observed that Native Americans of the Upper Missouri River region used *Echinacea* for more ailments than any other plant, including frequent use as an analgesic and local anesthetic. The Blackfoot and Lakota used it for toothaches. Other tribes used *Echinacea* to treat snakebites and other venomous bites, stings, and poisonings; putrefied wounds; hydrophobia; inflammation; burns; sore eyes; toothaches; tonsillitis; stomachache; and bowel pain. Daniel Moerman's database of Native American ethnobotany lists 119 documented applications of *Echinacea angustifolia* and *Echinacea purpurea* drawn from historical ethnography, including studies of Blackfoot, Montana, Omaha, Lakota, Pawnee, Ponca, Sioux, Winnebago, Kiowa, Meskwaki, Cheyenne, Crow, Dakota, Comanche, Delaware, and Choctaw communities.[7]

Lloyd's efforts were not wholly novel: the plant's uses had been documented in a number of European and American works on medical botany before Lloyd turned his attention to it.[8] Rather, his efforts came at a time of transition for medical and pharmaceutical practice in the United States. A main staple of regular practice for centuries, botanic medicine in the United States was soon to be derided as the province of hucksters and quacks.

Lloyd's attempts to acquire a supply of *Echinacea angustifolia* for a single year's manufacture in 1903 illuminates changes in labor, land use, environment, and medical knowledge at the turn of the century. Lloyd Brothers' branded production of Echafolta, a purified, assayed form of *Echinacea*, required thousands of pounds of the plant. This production was not by any stretch of the imagination traditional, nor did his faltering attempts to secure raw materials indicate the passing of an age-old art. Rather, his business attempted to translate certain traditions of botanic medical practice into mass production, with some difficulty. These difficulties, and what they reveal about material and intellectual resources on the ground, illuminate the marginalization of botanic medicine at the turn of the century, as large-scale agricultural development altered ecosystems and lay knowledge about domestic plants.

The racist, ambivalent, and multivalent vernaculars for *Echinacea* suggest the heterogeneity of knowledge about indigenous plants at the turn of the century. In the eighteenth- and nineteenth-century United States, diverse people and institutions managed what we would now refer to as drug development. These included compounding druggists, country physicians, farmers, local healers, midwives, and plant collectors of Euro-American, indigenous American, and African descent. They drew on varied legacies of botanical knowledge and medical practices associated with them.

These healers came into conflict with competing claims on natural resources as more land was brought under cultivation: a contest of definition expressed by the many names applied to the plant. Some regarded *Echinacea* as a medicine, others as a weed. The latter constituency carried the day, ultimately eroding diverse sources of local knowledge about plants and their powers. Although typically explained as the triumph of professionalizing physicians and progressive reformers at the turn of the twentieth century,[9] the marginalization of botanic medicine in the United States was as much a result of changes in land use and political economy as of medical knowledge.

John Uri Lloyd's early professional activity provides a window into the practice of pharmacy in the United States in the decades after the Civil War. Lloyd was born in 1849 in West Bloomfield, New York, moving to Kentucky four years later, when his father took a job as a railroad surveyor. Construction faltered in years of depression, and both Lloyd's parents became schoolteachers, serving the growing settlements in Kentucky and neighboring Cincinnati. Before Cincinnati became the "Porkopolis," it was "Queen City," strategically located to move goods down the Ohio River. From 1825, German and Irish immigration surged, along with a small population of African Americans from Kentucky and Virginia. By 1850, it was a city with 278 doctors, 153 druggists, and four medical colleges.[10]

Lloyd entered the profession of pharmacy through a series of apprenticeships, the most common path to a career in the early and mid-nineteenth century. During the Civil War, Lloyd's father sought a placement for him in a Cincinnati drugstore. He joined the business of William J. M. Gordon and Brothers in the fall of 1863 and served there two years, the first of his fourteen as a compounding pharmacist. After his tenure with Gordon, Lloyd served another two years as an apprentice for George Eger, a German-trained pharmacist. While there he attended chemistry lectures at the Medical College of Ohio before taking a position as prescription clerk at Gordon's. He swept floors and delivered orders and managed, in the process, to learn a few things.[11]

Gordon was a prominent figure in American pharmacy at the time, founding the first college of pharmacy west of Appalachia above his shop in 1850 and serving as president of the American Pharmacists Association in 1864. Gordon's position conveys the rapid growth of the industry during the boom-and-bust years of western settlement, as well as the emerging national orientation and self-regulation of pharmacy by the late 1850s. Beginning in the 1870s, municipal licensing acts proliferated throughout the United States. Lloyd passed Cincinnati's licensing test in 1873, the year it was inaugurated.[12]

Lloyd's additional apprenticeship with George Eger reflects the influence of German pharmacy in the United States, especially in the western states. Immigrant communities brought medical traditions with them across the Atlantic. Although histories have prioritized organized movements such as Samuel

Figure 6.1. "Blacksamson Echinacea" collected in the Upper Platte area on John C. Frémont's third western expedition (1845–47). US National Herbarium, National Museum of Natural History, Smithsonian Institution.

Hahnemann's Philadelphia school of homeopathy, the influence of German botanic traditions was not confined to institutionalized practice. When Christian Krehbiel nursed a sick cousin, the latter asked his doctor, who was also of German ancestry, whether he could give him anything other than the filthy water and medicine he was receiving. "Alas, no!" the doctor replied. "In Germany I could give you dry plums, raspberry juice, and other prescriptions, but here in the West we must do with what we have." In spite of limits on the availability of supplies, German botanic traditions filtered through a wide array of unaffiliated pharmacists', physicians', and home healers' practices.[13]

Nor was botanic practice principally German. Plant drugs of global origin dominated western European pharmacopoeia through the nineteenth century.[14] Their prominence was an artifact of European imperial exploration and the global trade in commodities it supported. During and after the sixteenth century, a transatlantic drug trade linked the West Indies, Spain, and the Americas. Ultimately, Dutch, British, and French explorers fueled the quest for New World plants of economic value, including dyewood, sugar, tobacco, cinchona bark, sarsaparilla root, sassafras, and China root. In putting nature to the service of empire, explorers and merchants dramatically expanded the corpus of the prevailing herbal medicine traditions.[15]

Meanwhile, efforts to publish an American pharmacopoeia to rival London and Edinburgh's own culminated in an 1820 convention in Philadelphia, followed by the publication of a national formulary.[16] A few elite physicians, including Benjamin Smith Barton, expressed interest in Native American materia medica, especially as it served a nationalist project. Others, including the prominent physician Benjamin Rush, considered Native American medical practices rudimentary, if, as Rush argued, suited to the simplicity of the people. A proponent of so-called heroic medicine, Rush advocated strong interventions, especially bleeding afflicted patients. When Rush studied the practices of New England Indians, he focused on these aspects of their practice without noting as others had the wide variety of plant medicines they prepared.[17]

For urban physicians such as Benjamin Rush, as for company surgeons, the vast majority of drugs came to American harbors from London, the center of the international drug trade since the seventeenth century. The principal New England apothecaries dealt directly with drug brokers in London. Well into the nineteenth century, most of the world's supply of raw drugs was sold at auction at Mincing Lane in London, the heir of seventeenth-century Navigation Acts and British imperial expansion. A Warehousing Act of 1803 created an infrastructure of dock police and bonded warehouses where samples were classified and prepared for auction, rationalizing the drug trade from production to point of sale. While the scope of markets expanded, London remained the hub and conduit for market information, reflecting the city's status as the center of international shipping, finance, and insurance. At least through the 1870s, London remained the center of the trade, controlling both the flow of materials and information about them.[18] By the mid-nineteenth century, drug importers had set up shop in New York and Philadelphia, drawing on this established international trade.[19]

Typical medicine chests reflected the botanic character of eighteenth-century medicine, as well as its Old and New World influences. When fishermen, fur traders, and other early settlers had access to medical care, it was usually from a company surgeon with modest supplies. Surgeon Giles Wills's 1730 inventory for the Hudson's Bay Company outpost at Fort Albany, New York, included wax, pitch, and turpentine, for ointments; mustard seed, pepper, verdigris, and cantharides; tobacco (cleansing); fennel (for digestion); marshmallow root (for inflammation); Guaiac (as a diaphoretic, for fevers and syphilis); and cinchona (Peruvian bark, for malaria).[20]

In spite of well-developed institutions of international trade, a tradition of self-care inaugurated by early European settlers characterized Euro-American medicine. Through the nineteenth century, plant drugs formed the principal stock of most healers, regardless of tradition. The panniers of any country physician would be stocked with botanic preparations, as would the big house on a Piedmont plantation, the quarters of an African American midwife, and the medicine chest of the surgeon aboard a naval expedition to the South Seas.[21] Especially in the Appalachian region, a commercial culture of botanic medicine thrived, with herbalists gathering roots and flowers for the use of new northeastern pharmacies, local apothecaries, families, and healers.[22]

For most nineteenth-century Americans, moreover, home health care based on botanic practice was the norm. Popular works on home health care proliferated, including John Tennant's *Every Man His Own Doctor* and William Buchan's *Domestic Medicine,* the latter issued in over a hundred editions after its first publication in 1769.[23] By the 1830s, these were joined by John Gunn's *Domestic Medicine; or, Poor Man's Friend,* which enjoyed a similarly lengthy publication run. The title page of Gunn's guide noted that it was "expressly written for the benefit of families in the Western and Southern States" and contained "descriptions of the medicinal roots and herbs of the Western and Southern country, and how they are to be used in the cure of Diseases."[24] Each of these guides contained a lengthy catalog of recipes for common medicines, many of which relied on locally available plants.

The commercialization of botanic preparations proceeded apace with geographic expansion, catering to western settlers without access to urban physicians and medicaments. As towns became cities, compounders of botanic medicines began to package and sell their wares to wider markets, building on a long tradition of English and continental European patent medicines.[25] Swaim's Panacea, for example, first applied to treat syphilis, was an American variant of the eighteenth-century French proprietary medicine, the Rob d'Laffecteur. The Rob d'Laffecteur, in turn, was a preparation of sarsaparilla, the Central and South American root Spanish explorers found natives using for venereal afflictions. While pharmacists continued to stock generic sarsaparilla preparations, William Swaim peddled his secret nostrum with such flair that it became a runaway bestseller for all who ailed. His secret potion, widely known to be a concoction of mercury and sarsaparilla, inspired numerous copycats—and much criticism from physicians and druggists who knew what he was up to.[26]

Whether or not they shared Swaim's marketing prowess, many producers sought to make legitimate and illegitimate preparations available to a wider market.[27] Lydia Pinkham combined several indigenous medicinal plants in her famous vegetable compound and remedy for female complaints. These included black cohosh (*Cimicifuga racemosa*), indicated for a range of reproductive issues. Black cohosh entered the US pharmacopoeia in 1830 as "black snakeroot." Eclectic physician John King popularized it around the same time for the treatment of rheumatism and nervous disorders. Pinkham's compound also included butterfly weed (*Asclepias tuberosa*), or pleurisy root, which Native American women sometimes consumed after childbirth, as well as unicorn root (*Aletris farinosa*) and golden ragwort (*Packera aurea*, formerly *Senecio aureus*), the former used to prevent miscarriage, and the latter used to prevent pregnancies and childbirth complications. In large doses, all three were also suggested as abortifacients. Cohosh and unicorn root have estrogenic effects. Ragwort's efficacy derives from toxic pyrollizidine alkaloids. Pinkham advertised her compound, macerated and preserved in alcohol, for female troubles including depression, debility, prolapsed uterus, and tumors of the uterus.[28]

Soon country stores and corner druggists stocked scores of branded tonics, infusions, and sarsaparillas such as Meyer's, Pinkham's, and Swaim's. This motley marketplace of imported European drugs, native botanicals, and packaged over-the-counter remedies was typical for nineteenth-century practice. Only in the later nineteenth century did the large drug houses such as Parke-Davis, Wyeth, and Squibb come to prominence. Notably, Wyeth and Squibb benefited from sales to the Union army, putting them in company with many industries from meatpacking to apparel that owe their success to the Civil War.[29]

As the realm of pharmacy diversified and expanded, so too did the theory and practice of botanic medicine. By the 1830s, many proponents of botanic remedies organized into semiofficial sects.[30] Beginning in the 1820s, many practitioners had rejected harsh therapies such as calomel (mercury) and bloodletting. One of the earliest and most prominent critics was Samuel Thomson, who published his therapies in *The New Guide to Health* (1829).[31] Thomson organized many branches of followers in the southern and western states. He professed the body's ability to rebalance itself with the aid of mild botanic agents, sweating, and expulsion. There were considerable differences between movements, in spite of which many lumped all botanic sects together, deriding the whole lot as "steamers and pukers."

The effort to treat sick and wounded soldiers during the Civil War exposed the diversity of practices among schools, botanic sects, and individual healers. The war brought together diverse traditions of healing on the battlefield, but the medical wagons, panniers, and knapsacks of Union medics stocked only the most widely applied remedies. The panniers of a Union medic were packed with opium, laudanum, and mercury produced by Squibb and other manufacturers, as well as by the US Army Laboratories during a brief experiment in nationalizing production to offset price spikes and shortages in the supply of essential drugs. Yet numerous medics complained about lack of access to the botanic supplies with which they were most familiar. Medics came from diverse backgrounds, including employment as retail pharmacists, country physicians, and hospital stewards. Enlisted men with no formal training might also serve medical roles in the military.[32]

Historians of medicine have argued that the rambunctious democracy of nineteenth-century medical sects delayed the consolidation of medical authority in the hands of allopathic physicians until the early twentieth century.[33] Gradually these physicians, who advocated drugs that counteracted symptoms, gained the upper hand. Their success was not preordained, however, nor was their authority a solitary cause of the marginalization of botanic medicine. Botanic sects were legitimate options among an array of medical philosophies competing for authority, not alternatives to a single normative practice.[34]

In fact, there was an extraordinary degree of continuity between healers of all persuasions during the nineteenth century. Differences in theory rarely translated into practice. Rush, Thomson, and the vast majority of American practitioners in the eighteenth and early to mid-nineteenth centuries shared the belief that the body was a self-regulating entity that required balance, however it might be achieved. All classified drugs according to their effects on the body, rather than in reference to specific diseases. Whether they preferred chemical remedies like mercury to botanicals like sarsaparilla in the treatment of syphilis, for example, they valued therapies for their ability to produce symptoms in the body, not to suppress them. Healers shared the practice of "exhibiting" a drug: administering it to the patient and witnessing its effects on the body, such as salivation, vomiting, and sweating. Although the nineteenth-century sects had articulated intellectual foundations, in practice they were quite similar to the therapeutics of any practicing physician in privileging emetics, cathartics, diuretics, diaphoretics, and expectorants.[35] But their preferred medicines and methods might differ considerably.

Lloyd was still a boy in the war years, during which he trained under compounding pharmacists in Cincinnati with no particular sectarian allegiance. In 1871, he made the professional decision to serve as the compounding pharmacist for John King, the preeminent Eclectic physician. Thereafter, his practice was strongly associated with Eclectic medicine. In 1876, he became a partner in the firm Merrell, Thorp, and Lloyd, and in 1882 he partnered with his brothers Ashley and Curtis to found Lloyd Brothers Pharmacy. Brother Curtis took a special interest in the identification and study of indigenous plants, including *Echinacea*. Lloyd Brothers sold herbs as simple preparations and branded Specifics, including Echafolta.

The leaders of the Eclectic movement diverged from Thomson and others in advocating orthodox medical education. Under the direction of John King and Wooster Beach, colleges and universities associated with Eclectics flourished. These institutions became training grounds for physicians from small towns, villages, and farmlands. Once trained, graduates returned home to practice. The Eclectic Medical Institute in Cincinnati also admitted students often barred from education at other institutions, including women, Orthodox Jews, and African Americans. Whether diversity was a conscious enrollment policy or an inclusiveness born of economic necessity, the Eclectic Medical Institute trained a broader population of physicians than the average medical school of its era.[36] The Eclectics earned their moniker by taking what was around them and adapting it to medical use, including commercial and indigenous products. More so than the other botanic movements, they prized Native American knowledge of medicinal plants and indigenous materia medica in general.

Eclectic openness extended to proprietary medicines, and it was through this path rather than ethnobotanical texts on indigenous pharmacy that *Echinacea* came to the attention of John King. By the 1880s, H. C. F. Meyer of Pawnee City, Nebraska, claimed to have cured 623 rattlesnake bites with the aid of the plant, which he bottled and marketed under his own name. Meyer, who had devoted considerable energy to capitalizing on *Echinacea*'s virtues, recommended it enthusiastically to John King in hopes that he would promote its reputation among the medical community. King did so cautiously. "Dr. Meyer," he wrote, "entertains a very exalted idea of his discovery, which certainly merits a careful investigation by our practitioners . . . and should it be found to contain only one-half the virtues he attributes to it, it will form an important addition to our materia medica." John Uri Lloyd was even more skeptical than King when approached about bringing an *Echinacea* tincture to market. Meyer wrote to Lloyd suggesting a visit to Cincinnati with his usual trunk show, a production involving subjecting himself to multiple rattlesnake bites and applying his patented *Echinacea* formula. Lloyd declined.[37]

Yet when King did persuade Lloyd to produce *Echinacea* in 1889, there was already a large demand among physicians, in part stimulated by Meyer's popularity. Lloyd's release further increased demand. Echafolta and Libradol quickly became the company's biggest sellers and a focus of its marketing efforts.[38] These brands reflected the Eclectic emphasis on Native American drugs and medicines. Lloyd compounded Echafolta from *Echinacea angustifolia* (purple coneflower), and Libradol from *Lobelia inflata* (Indian tobacco) using domestic supplies, and they were indicated for afflictions nearly identical to those treated by the herbal medicines from which they derived.

Lloyd's products were not proprietary medicines in the classic sense, in that they were not marketed directly to consumers. Rather, they were to be prescribed by physicians and dispensed by pharmacists, albeit largely Eclectic ones. In the polarized practice of the later nineteenth century, this rendered Lloyd's drugs "ethical" medicines rather than "patent" ones, separating him from Meyer's unscrupulous advertising stunts. Yet because Lloyd's products represented botanic practice dominated by proprietary and folk medicines, they fit awkwardly into the emerging division between ethical and commercial medicine.

Neither were Lloyd's products simple appropriations of indigenous medicine. Rather, Lloyd transformed Native American materia medica through the application of European alkaloid chemistry. In addition to his work on behalf of the Eclectic movement, Lloyd was a noted chemist. He received prizes for his work standardizing assays for alkaloid products and for his wider research on fluidextracts, considered fundamental to the development of colloidal chemistry. His laboratory notebooks are a testament to his energy as a practicing chemist and his methodical experiments in alkaloid assays.

In addition to Echafolta and Libradol, Lloyd focused his energy on the research and development of new branded "Specifics": liquid preparations containing the active constituents of a single plant. The 379 he developed were widely used by both Eclectic and non-Eclectic physicians, their quality universally respected. Some of the apparatus Lloyd developed in conjunction with the Specifics, chiefly the cold still extractor, is still in production.[39] But when Lloyd attempted to source *Echinacea* for the manufacture of Echafolta, he had to negotiate varied economies of knowledge and changes in the land.

Lloyd was catholic in his approach to people and plants, having served for years as a professor of chemistry at a medical institute staffing the front lines of rural medical care. He was also a businessman, determined to secure a supply of *Lobelia* seed for Libradol and *Echinacea* for Echafolta at the lowest possible prices.[40] By the time Lloyd began producing his *Echinacea* tincture, many believed the prairie lands of Kansas and Nebraska produced the best roots of *angustifolia*, and this is where Lloyd sought his supplies.[41]

When he set about acquiring a supply for the upcoming season, he did so through multiple channels. He printed a notice in the *Medical Outlook*, a subscription publication for Eclectic physicians and other interested parties. He distributed a circular to his extended correspondence network, consisting of known suppliers, local collectors, and physicians. He made inquiries of agricultural colleges and experiment stations in many states in the South and West. And he inquired directly of regional dealers in medicinal plants and the major drug importers in New York, Philadelphia, and Baltimore.

When it came to acquiring *Lobelia* seed, Lloyd approached an established network of regional dealers in Jacksonville, Florida, and the Appalachian region of North Carolina. While Jacksonville dealers universally reported having no supply of *Lobelia* herb or seed (surprising, since it grows there in abundance), nearly all the North Carolina dealers had some stock of one or the other. He also tried to ascertain the plant's abundance in Wisconsin, but almost universally, agricultural experiment stations, colleges, physicians, and individual collectors reported the plant scarce. The South was *Lobelia*'s primary habitat, and a robust network of dealers collected plants there.

These dealers rarely dealt exclusively in medicinal plants. More often, their letterhead indicates their primary business as furniture dealers, dry goods merchants, or fiber traders. F. P. McGuire, Greer, and Co. had two versions of its letterhead, one touting their capabilities as "Dealers in General Merchandise and Country Produce," the other their activities as "Dealers in High Grade American Botanic Drugs, Essential Oils, and Ginseng."

In North Carolina, the market was small enough that Lloyd's large orders could skew prices dramatically. The Wallace Brothers Company for Roots,

Herbs, and Barks in Statesville, North Carolina, reported that they had booked Lloyd's order for 1,000 pounds of herb and 4,000 pounds of seed. They were busy collecting the latter, as they had only 1,000 pounds or so on hand. They reported that Lloyd's inquiries "to other parties some time since and as well as recently has had a tendency to advance price to collectors, and has caused us to lose several lots." Of course, it was in their best interests to make this claim. But there was at least a grain of truth to it: McGuire, Greer, and Co. in particular at first told Lloyd their stock was closed out, and then apparently scrambled to renew it, writing him twice more with updated rates.

Firms like Wallace Brothers and McGuire, Greer, and Co. contracted collectors to do the work of gathering the plant. Nevertheless, the best of them could specify the quality and circumstances of collection. J. Q. McGuire of Asheville, North Carolina, "Wholesale Dealer in Medicinal Roots and Herbs, Ginseng, Beeswax, Essential Oils," quoted Lloyd prices for the collection of Lobelia Herb and Seed, specifying that it was "common quality and gathered after seeding."

We know little about the collectors themselves. In all likelihood, they were a mixed lot of agricultural laborers with skilled knowledge, sons of country physicians, local healers, and jacks-of-all-trades. Contemporary ethnographies of Appalachian herbalists give some sense of the oral traditions and long histories associated with local knowledge of medicinal plants, especially in mountainous regions remote from interregional and national markets.[42]

For both *Echinacea* and *Lobelia*, the large drug importers in Eastern Seaboard cities had ready supplies. Davis and Davis of Baltimore, J. L. Hopkins, Lehn and Fink, and Parke-Davis in New York reported stock. In New York, the major drug importers clustered near William and Gold Streets. Although sufficient company records do not survive to verify the claim, Lloyd appears to have had a strong preference for using locally and domestically gathered supplies, which afforded him more control over price, quality, and collection.

Unlike *Lobelia*, *Echinacea* posed special challenges to procurement. It grew wild only west of the Mississippi, where there were no established networks of collection for many medicinal plants. Lloyd did procure a sample of some flowers from A. W. Krauss and Co. of Jacksonville, Florida, which dealt in medicinal plants, but according to the USDA botanists Lloyd consulted on the matter, the plant was *Helianthus radula* rather than *Echinacea angustifolia*. While country merchants organized dispersed collectors in Appalachia, no such businesses existed west of the Mississippi, where Lloyd had to rely on his own correspondence network and that of the Eclectic Medical Institute.

County health officials might provide some support. W. F. Flack, MD, the

Elk County health officer in Longton, Kansas, enclosed a sample of the root for Lloyd's inspection, indicating that he could supply from 200 to 500 pounds, and recommending a man by the name of John Rupp as a collector.

Where they failed, he looked to the emerging agricultural colleges and experiment stations to fill the gap. His correspondence with USDA botanists provides further evidence of his reliance on agricultural research and extension services. But these institutions were oriented toward agricultural production, and especially large-scale grain culture, not domestic medicine. Pharmacy departments of land grant universities provided some resources. In 1897, Dr. L. E. Sayre of the University of Kansas Pharmacy Department noted the use of student labor to collect plants during the late summer and early fall months, during which they might "find in it a little profit at twenty-five cents a pound." Interest in the plant was sufficient to motivate chemists at the University of Kansas to isolate the oil of the root in 1898.[43]

Lloyd pursued his supply of *Echinacea* at the high-water mark of demand for the plant, a testament to the success of Meyer's advertising campaign and the subsequent success of Echtafol in the several years since its release. Dr. L. E. Sayre of the University of Kansas Pharmacy Department reported that over 200,00 pounds of dried root worth over $100,000 were harvested in 1902, during which prices rose to fifty cents per pound from twenty-five cents per pound five years before. A pound consisted of eight to ten roots, so roughly two million roots were harvested in 1902, the year before Lloyd sought supplies.

When Lloyd searched for *Echinacea,* he deployed official and vernacular names and requested the help of readers in securing a supply. When he wrote to dealers, he used the plant's Latin binomial: *Echinacea angustifolia.* In his circular in the *Medical Outlook,* he also identified the plant by its vernacular pejorative, "Nigger Head." He added that the root occurred in most sections of Kansas, and that he hoped readers "could suggest some party who would gather it."

Lloyd received a variety of responses, most addressed to him personally. One correspondent, the secretary of the Eclectic Medical University in Kansas City, Missouri, could not believe the ill-smelling and rank weed he knew was the great medicine *Echinacea.* He nevertheless agreed to find and gather some personally—but presumably not the thousands of pounds Lloyd desired. Another correspondent reported that his father, an old stockman near Longton, Kansas, had considerable quantities growing in his fields.

A Dr. Baldwin in Waco, Texas, a former student of the Eclectic Medical Institute in Cincinnati, wondered if another "'yaller' headed flower" was the same

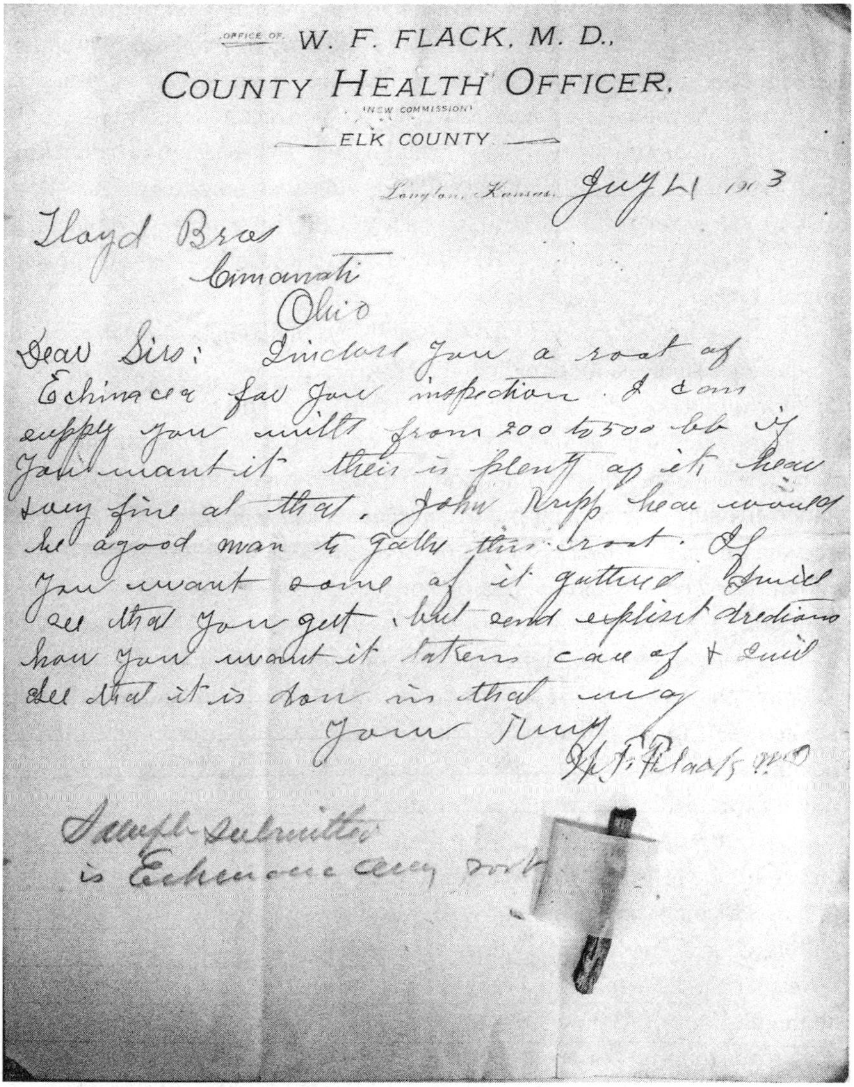

OFFICE OF W. F. FLACK, M. D.,
COUNTY HEALTH OFFICER.
(NEW COMMISSION)
ELK COUNTY

Langton, Kansas. July 21 1903

Lloyd Bros
Cincinnati
Ohio

Dear Sirs: Inclose you a root of Echinacea for your inspection. I can supply you with from 200 to 500 lb if you want it there is plenty of it here very fine at that. John Krupp here would be a good man to gather this root. If you want some of it gathered I will see that you get, but send explisit directions how you want it taken care of & I will see that it is done in that way

Yours Truly
W. F. Flack M.D.

Sample submitted
is Echinacea [illegible] sort

Figure 6.2. W. F. Flack, the Elk County health officer in Langton, Kansas, forwarded Lloyd a sample of *Echinacea* and recommended a local man to gather it. Lloyd Library and Museum, Cincinnati, Ohio.

species. Baldwin transcribed the Texas twang in self-parody and humility. He wrote that he got most of his "yaller harb" from a rocky point at the junction of the Brazos and Bosque Rivers. He described how he used it in his practice, loosely packing the yellow flowers, covering them in alcohol, and drawing off a preparation quite similar to Lloyd's "Ecafolta [*sic*]." He also sunned batches for three days until they turned black, and then squeezed out the oil. "Now dont you go to New York and tell Merk [*sic*] about this," he admonished playfully. "I am so darned ignorant that I'll be dinged if I know whether I have Echinacea or Ecafolta or not."

Other responses directed Lloyd to established plant collectors. Lloyd's correspondence suggests that medicinal plant collecting was a specialized, if low paid, form of labor. The names of several dedicated and practiced collectors appear repeatedly in responses to Lloyd's circular. Several mentioned Terry Tharp, a "long time western practitioner" in Cherryvale, Kansas, who had also trained his son in gathering medicinal plants. Tharp Jr. was recommended as "a reliable agent." Jack Fishburn of Medicine Lodge, wrote another, "does that kind of work." Fishburn's name appears several times in Lloyd's correspondence. He appears to have been a reasonably successful businessman for whom collecting may have remained a secondary employment.

A physician in Paola, Kansas, referred Lloyd to "a colored man who digs for me each year" and testified to his reliability. Lloyd then corresponded directly with the man, Bud Payton, about securing a supply of *Echinacea* in Kansas. Payton expressed concern that rains and scarcity would prevent collection of any size but requested a price per pound. Payton lacked ready access to stationery and ink. His letter from the Paola post office survives as a faded photocopy of a disintegrating sheet of wide-ruled paper, in pencil. The script and spelling suggest an elementary level of education.

Nevertheless, Payton had a little Latin, or at least enough to correspond in the lingua franca of Linnaean binomials. When he wrote, he used the Latin name for *Echinacea* but erred in calling it "angustifolius" rather than "angustifolia." It was minor error that would not have mattered to Lloyd. Rather, it establishes that Payton had used enough Latin in reference to plants that he could mistakenly use the masculine rather than the feminine suffix for the subspecies. More significantly, Payton's use of Mönch's Latin for "hedgehog" points to the failures of the vernacular classification of medicinal plants Lloyd was seeking.

What would Payton have called the plant if he weren't addressing Lloyd? Payton was born in the border state of Missouri in September 1860, on the eve of the Civil War, to parents from Virginia. It is likely that the war guaranteed his legal freedom shortly after his second birthday. He owned a house in Paola and married at the age of thirty-one. The 1900 census lists him as a teamster (a driver of teams of animals), which suggests that agricultural labor was his primary employment.[44] It is doubtful that Payton would have used either of the racist monikers applied to *Echinacea*. In fact, it's noteworthy that Lloyd initially estimated the most common name for the *Echinacea* imagining a population of white farmers. This address, however targeted, was both demographically inaccurate and bound to miss some of the most knowledgeable workers in the field.

Given that Payton was an agricultural worker, he might have known *Echinacea* as a weed. But it is also possible that Payton acquired additional training in medicinal plants, either from the Paola physician or from other healers in his community. *Echinacea* was in the repertoire of many different healers, including those caring for enslaved and free African American communities. One description of a Mississippi conjure man's hand noted the presence of "Samson's snakeroot," although it is not clear whether this is the conjurer's term or the observer's.[45] Payton may also have learned from Plains Indians settled in the region. Perhaps "Elk's Root" may have been its primary designation. Regardless, the vernacular lexicon for local flora wasn't Payton's, even though he had more knowledge of its environmental conditions than most people.

These minor errors in spelling—of Ecafolta, angustifolius, or yaller or harb—indicate a keen awareness of the political economy of pharmaceutical knowledge at the turn of the century. Dr. Baldwin of Waco, Texas, had a flower he wasn't sure was *Echinacea*, which he mockingly called the Yaller Harb after the twang of his adopted home. He apologized for his ignorance, nevertheless playfully warning Lloyd not to tell Merck how he prepared the flower's oil. The German pharmaceutical company, based in Darmstadt, had set up shop on Wall Street in 1887. Bayer, the German firm owning patents and trademarks for aspirin (a willow bark derivative), set up an American subsidiary in Rensselaer, New York, in 1903.

Baldwin was critical of physicians "too lazy to make up any medicines," an apology for independence and general training in the very moment physicians' and pharmacists' practices were diverging as specialties. At the same time, he regretted his own ignorance, declaring the Eclectic education too narrow to have educated him on botanic variations like this one. "Not when I was at School," he allowed. "(Nor now)" reads a pencil annotation in John Uri Lloyd's handwriting.

Nevertheless, he suspected somehow that his good practice, commodified, might be someone's best-selling product. Whether or not Baldwin had *Echinacea angustifolia*, he had something that he thought worked, or he would not have gone to the trouble of preparing it. He didn't begrudge his old professor anything. Nevertheless, he wondered if he was about to give something away that would be repackaged and sold back to him as a breakthrough medicine.

Meanwhile, Bud Payton had years of collecting experience and could probably recognize a wide array of medicinal plants—by means of whose training, we can't know. But to get the contract, he had to write in Latin.

Lloyd appears to have been successful in acquiring *Echinacea* from Kansas—although there were problems filling at least one of his contracts for 2,000 pounds due to unreliable labor, according to the contracting party. Many reported that it was indeed plentiful in their neighborhoods. *Echinacea angustifolia* and *E. pallida* entered the fourth edition of the *National Formulary* in 1916, remaining through four editions, until 1947. By 1917, Lloyd claimed his tincture was a therapeutic favorite with many thousands of physicians of all schools of medicine, consumed in greater quantities than any American drug in the previous thirty years.[46]

But Lloyd encountered numerous problems securing a supply in Kansas. In part, land use patterns had changed. As more land was cultivated for grain, many sites of collection were eliminated. Some noted that whereas the plant had been plentiful in the past, most of the land was now fenced or under cultivation. Additionally, pasturing and annual mowing seemed to "kill it out." One farmer reported that it grew plentifully in his cornfield, but that harvesting removed the plant before it was ready. Others declared the plant entirely eradicated by cultivation.

Permission to collect posed another problem. In fenced pasture, *Echinacea* might grow, but in addition to the costs of labor, Lloyd would have to pay for the right to dig roots. Most farmers would not allow it.

Several writers directed Lloyd to inquire farther west, where there was uncultivated prairie. Other correspondents had heard of the weed growing in Nebraska and Wyoming, including one correspondent from Reading, Kansas, who scribbled a note on the outside of the envelope about an instructive conversation he had at the post office as he was mailing the letter.

The environment in which *Echinacea* flourished also impeded economical collection. *Echinacea* grew plentifully on rocky precipices and in disturbed and remote areas, but then labor costs became prohibitive, creeping upwards of $1.50 per pound. Both reaching the site and breaking the earth with a pick were onerous tasks. Even the undisturbed prairie west of the cultivated areas had to be broken with a pick, the soil too stiff to yield to a spade or hoe. After all, it was in part the stiffness of the prairie soil that had stalled cultivation for so long.

The availability of labor was a problem, regardless of site. Lloyd contracted his supply around the time of harvest, when the flowers were likely to be mature, but most laborers were taken for the harvest and unwilling to divert energy to collecting flowers, unless they received high rates of compensation. Reportedly, others had been diverted to an oil and gas field nearby. In the end, Lloyd conducted a lot of business with drug importers in New York and Philadelphia.

The botanic medicines for which Lloyd's business was admired soon disappeared from the pharmacopoeia, destined to be rebranded as soda pops, redefined as foodstuffs, or classified as spurious nostrums or cosmetics. The Pure Food and Drug Act, passed into law in 1906 after years of agitation by allopathic physicians and muckraking reformers, took aim at herbal tonics and elixirs, especially those spiked with alcohol, opium, or cocaine. It required producers to list eleven "dangerous ingredients" (narcotics and barbiturates) on the label of the bottle, enabled the seizure of "adulterated" and "misbranded" drugs, and forbade "false and misleading" statements on labels. Meanwhile, the American Medical Association's Council on Pharmacy and Chemistry (1905) aimed to control drug quality through the award of a seal for drugs screened for purity and efficacy. In this emerging regime, "ethical medicines" predominated, so called because they required the prescription of a physician.[47]

While Lloyd's products were largely for prescription, the collateral disqualification of botanic remedies and Eclectic medicine damaged his reputation. Abraham Flexner's 1910 report *Medical Education in the United States and Canada* portrayed most medical colleges in the United States as woefully behind the cutting-edge clinical research of Johns Hopkins University in Baltimore.[48] The Eclectic Medical Institute in Cincinnati, for which Lloyd taught chemistry, was among the schools not to make the grade.

Meanwhile, the locus of drug development was shifting to the German chemist Paul Ehrlich's fantasies of a "magic bullet," chemical compounds that would act on disease while leaving the body untouched: the antithesis of the traditional theory in which the proof of a drug's efficacy was its ability to produce demonstrable effects on the body. In 1909, the Council on Pharmacy and Chemistry of the American Medical Association published a report declaring *Echinacea* useless.[49] In 1910, the ninth decennial revision of the US Pharmacopeia contained a minority of botanical drugs for the first time in history. While certain plant medicines such as aspirin (a dyestuff by-product derived from willow bark) became principal therapies, many more waned. Substances suited to alkaloid chemistry such as opium (morphine) and cinchona (quinine) remained prominent, while many more imperfectly understood therapies were neglected.

The tenacity of quackery and fraud as the primary characteristics ascribed to nineteenth-century American medicine is evidence of the success of muckraking reformers. To date, historians of medicine have described nineteenth-century medical sects in political terms, as evidence of rambunctious if democratic tendencies that hampered professional consolidation of allopathic physicians. In this version of events, reformers and good doctors won; quacks and hucksters lost. Reformed pharmacists allied with allopathic physicians purged the pharmacopoeia of useless weeds, elevating drug therapies that worked by counteracting symptoms of disease. In the process, they disqualified other ways of understanding bodily sickness and health.

The effect of these accounts has been to relinquish the writing of history to the professionalizers themselves, who sought to discredit diverse practices of medicine in the nineteenth-century United States. These accounts have distorted the history of medical and pharmacological knowledge by misrepresenting the diversity and legitimacy of medical practice its authors displaced: a history that includes not only the botanic sects of the nineteenth century and their commercial expressions, but also the numerous practices on which they drew.

Stories of fraud and reform have also minimized other explanations for the marginalization of botanic medicine, including environmental pressures. During peak periods, *Echinacea* was the most extensively harvested medicinal plant on the prairie, leading to periodic calls for conservation and/or commercial cultivation in the same moment the prairie itself was imperiled by large-scale grain cultivation. Wild stands of ginseng, goldenseal, and nerve roots were all eliminated by overharvesting during the nineteenth century. It is not clear that Lloyd's extensive manufacture of indigenous materia medica could have flourished in any event; for how many thousand pounds of *Echinacea angustifolia* could the vanishing prairies of Kansas and Nebraska yield?[50]

Lloyd decried the devaluation of plants and knowledge about them. In his writings, he turned from values of research and development to those of stewardship and history. Although he contributed to the industrial development of the United States, in late life he also became its critic. In 1902 he co-founded the Committee on Historical Pharmacy of the American Pharmaceutical Association, looking for the first time to traditions eclipsed by modern pharmaceutical and medical practice. Gradually, his interest in the future turned his attention to the past.

Lloyd's other writings, too, reflect an effort to link warnings about imperiled environments, types of knowledge, and ways of life. He wrote long essays about the natural history of Kentucky; the cliff dwellers of Canyon de Chelly in Arizona; his grandparents' flour mill in the little town of Factory Hollow; having to put down his old dog, Turk; the early settlement of Ohio; and the future of agriculture in America. He considered the country's pretensions to be "the world's granary" "vicious" and "extravagant." When he looked at the wheat belt, he saw "vandalism," not farming. He thought America's greatest gift, its soil, was being squandered. And he feared that the erosion of environmental complexity also meant a loss of knowledge derived from it.[51]

His anxieties were realized, inasmuch as overtly progressive histories of medicine have prevented us from considering how pharmacological knowledge might have developed differently had botanic practice remained prominent. Drug development was not a preordained progression from the chemical study of alkaloids to the fabrication of synthetic drugs. Nevertheless, by the 1920s the locus of development had shifted toward Ehrlich's magic bullet and the synthetic compounds associated with it. As a result, useful historical accounts of pharmaceutical practice tilt toward the postwar rise of the chemical industry as the birth of "Big Pharma": a history of the twentieth-century military-industrial complex obscuring the diverse traditions of local knowledge it transformed.[52]

In narrowing the field of knowledge from which drug development descended to the practice of laboratory chemistry and clinical research, histories have also obscured the attenuated yet persistent reliance on local knowledge about plants and their uses in formalized research and development. Disciplinary histories cordon off these researches as the province of phytochemistry and ethnobotany, giving both a narrow twentieth-century academic origin.[53] Yet albeit in a phytochemical frame, government laboratories, universities, and pharmaceutical companies have continued to look to plants as potential sources of new drugs.[54]

Lloyd's business changed hands several times after his death in 1936, first in 1938 to S. B. Penick, who continued the manufacture of its more successful lines, then in 1956, to Westerfield, and again in 1960, to the German pharmaceutical manufacturer, Hoechst, which relocated its operations.[55]

When Lloyd died, the Japanese pharmaceutical manufacturer Hajime Hoshi hosted a memorial service in Tokyo that drew over 1,000 people, prom-

Figure 6.3. ***Echinacea angustifolia*** **(purple coneflower) in the "Collection of Medicinal Plants," steppe meadow reserve, Stavropol Botanic Garden, Stavropol Krai, Russia. Photo by Courtney Fullilove.**

ising a gift of 5,000 cherry trees to the city of Cincinnati, where Lloyd Brothers Pharmacy was based.[56] (For comparison, the Japanese gift to Washington, DC, was 2,000 trees.) Unfortunately for the natural beauty of Cincinnati and John Uri Lloyd's claim to posterity, World War II interrupted the transit of the trees. And so the fourteen scientific books and some 5,000 scientific articles and editorials Lloyd produced in his lifetime survive mainly as arcana in the public domain on Google Books, and the drugs he developed were re-regulated as cosmetics and vice goods. And nobody remembers John Uri Lloyd, except, curiously, for his cult science fiction novel *Etidorhpa* ("Aphrodite" spelled backward), about an abducted chemist who journeys into the center of the earth through the mouth of a cave in Kentucky.[57]

By the time Lloyd died, he was already on the periphery of legitimate medicine in the United States. But Hajime Hoshi could celebrate him as a captain of industry, and the German firm Hoechst could acquire his declining company, because it was not so much useless as temporarily illegitimate in a local context. The prominence of Kampo medicine in Japan and herbal drugs in Germany provide a counterpoint, evidence that given a different set of visions for land use and bodily integrity, research and development in the United States might have taken another course, with indigenous materia medica at the center of practice.

Moreover, as the American experience suggests, botanic medicine was not a single tradition or theory of health. Euro-American botanic medicine is a hybrid of Atlantic traditions drawn from European, African, and indigenous American practice. Nor can we say that there was a single *Echinacea*, even distinguishing between subspecies *purpurea* or *pallida* or *angustifolia*: for in becoming Elk's Root, Kansas Snakeroot, or Black Samson, the plant metamorphosed into not one but many medicines. Lloyd's preparations drew on these traditions of therapy while attempting to remake them for a national marketplace. Ultimately, his knowledge and labors would become as obscure as Payton's, the conjure man's, or the elk's. The animal that may have first noticed that wildflowers helped with pain became extinct even before the wild grass prairie he inhabited.

7 : THE ALLEGORY OF THE CAVE IN KENTUCKY

Lloyd wasn't the only one who lamented the extinction of plants and animals on the plains. The elk, lamented Theodore Roosevelt, "is unfortunately one of those animals seemingly doomed to total destruction at no distant date. Its range has already shrunk to far less than one half its former size." Progressively eradicated from the Atlantic Seaboard and the eastern states, by the time Roosevelt wrote in 1885, the elk could only be found far west of the Mississippi, and then only in the deep mountain forests. Whereas formerly the elk was "plentiful all over the plains, coming down into them in great bands," "skin hunters" and "meat butchers" had waged a "relentless and unceasing war" for hides and flesh. Critical of trade hunters and cattlemen, Roosevelt advocated the elk's preservation for sport hunting. Roosevelt pursued the animals through the forest on foot with a rifle, rather than running on horseback across the plains as had been common when they were plentiful. In his view, sportsmen and lovers of nature could only regret the "gradual extermination of . . . the most stately and beautiful animal of the chase."[1]

Roosevelt and Lloyd were both advocates of preservation, but their styles and objectives differed. As Lloyd decried the vandalism of the prairie for monocultural production, Teddy Roosevelt called for the creation of national parks to safeguard nature's treasures—a federal and state enclosure movement that celebrated the scenic and the wild to the exclusion of other ways of inhabiting the environment. Conscripting rural land for parks prohibited its customary use for hunting and fishing. In some instances it also encouraged transitions from game hunting to livestock grazing in order to support the land claims of displaced inhabitants.[2] This movement devoted little attention to local knowledge of indigenous plants and their uses, such as interested John Uri Lloyd in his survey of the wild grass prairie.

Roosevelt's thirst for the hunt was inseparable from his patriotism, both radiating from his ideals of manliness. He believed hunting preserved the values of strength and courage required for healthy democracy. "Aggressive fighting for the right," he famously asserted, "is the noblest sport the world affords. . . . If I must choose between righteousness and peace, I choose righteousness."[3] And it was in this context that Roosevelt volunteered for the First United States Volunteer Cavalry, or the Rough Riders, a regiment raised to support Cuban independence during the Spanish-American War of 1898.

So it was that in the summer of 1898, while Teddy Roosevelt was charging up San Juan Hill, the commissioner of the Internal Revenue Service was trying to figure out how to fund the future president's exploits in Cuba and the Philippines. Several weeks before, Congress had passed a War Revenue Act taxing a variety of consumer articles, including alcohol, tobacco, perfumes, and cosmetics—and so-called patent or proprietary medicines. It was in connection with the last of these that Lloyd was summoned to Washington.

Lloyd was quite proud of the fact that in addition to being a famed author and a business leader, he was a "plain clothes man," or secret agent for the US government—or so he wrote in an essay he sealed to be published after his death.[4] Perhaps it's overgenerous to call a consultant to the Internal Revenue Service on a stamp tax a secret agent, but this was John Uri Lloyd's lot, and he knew it. Three years after the publication of *Etidorhpa,* a work of science fiction about a man who disappears into a cave in Kentucky into the center of the earth, Lloyd was called to Washington to advise which medicines were taxable under the new law. When he arrived, Lloyd encountered a badly conceived piece of legislation that dimly reflected provisional outcomes of infighting in the medical profession over who should have the authority to dispense medicines to patients. His task was to identify patent or proprietary medicines so that they could be taxed.

If his exploits were distinctly unromantic, the bureaucratic and legal minutiae they left behind nevertheless reveal immense confusion over what constituted a legitimate medicine at the turn of the twentieth century. Lloyd understood the contours of the debate enough to make its imagined classifications material. No stranger to the most far-fetched of fictions, Lloyd reluctantly played a role in his business's undoing by clarifying the logic of a tax that made little room for his own products, and created a zone of invisibility for the marketing of ethical drugs under the garb of industry standards and scientific vocabulary. His success as a bureaucrat made his fictional renderings of science all the more fantastical; in his efforts to make defensible rules for the IRS, he became ever more critical of the scientific certainties on which it relied.

Lloyd and Roosevelt were part of a broader movement for preservation at the turn of the century. Preservation and conservation were impulses that emerged in tandem with the agricultural and industrial development of the late nineteenth century. As defensive measures rather than formulated critiques, their objects often remained undefined: what precisely was to be preserved, and why? Both Roosevelt and Lloyd valued nature, but in different respects. For Roosevelt, its wilderness was essential to the cultivation of manliness. For Lloyd, its complexity was essential to the cultivation of humility, and this outlook led him to focus as much on knowledge systems as on land use as mechanisms for the organization of natural resources. It motivated his critique of science, and his elevation of stewardship and principled uncertainty as ways of living.

Lloyd's adventures at the IRS revealed bureaucratic disarray that made the Patent Office museum look like the Dewey decimal system. In his attempt to put the house in order, he consciously devised a classification as provisional and arbitrary as those Spencer Baird and George Brown Goode applied to the spoils of government science and military expeditions. There had to be rhyme or reason; there had to be a rationale for the War Revenue Tax.

As Lloyd tells it, when he got to Washington he found the IRS in an "inexpressibly deplorable condition," with Deputy Commissioner George Wilson distraught, allegedly fending off lawyers who were prepared to attack from every direction. He presided over a bureaucratic nightmare in which hundreds of clerks occupied a massive hall, buried in piles of arbitrarily sorted labels, haphazardly dividing things into stacks: taxable and nontaxable, for this reason or that.

According to Wilson, Congress had thrown the Stamp Tax Law on his office without any notice or instruction as to how to inaugurate it. While the clerks had done their best under the circumstances, they had "crossed themselves in their rulings" as to what was taxable on so many occasions that the IRS was "threatened with suits from all directions." In addition to which, a "mountain of untouched materials" was on hand to rule upon, "and thousands of labels coming from every direction, with requests for information as to whether the material described is taxable, or free."

Section 20 of the act specified which medicines were taxable with reference to packaging and claims. It read:

> The stamp taxes provided for in Schedule B of this act shall apply to all medicinal articles compounded by any formula, published or unpublished, which are put up in a style or manner similar to that of patent, trade-mark, or proprietary medicine in general, or which are advertised on the package or otherwise as remedies or specifics for any ailment, or having any special

claim to merit, or to any peculiar advantage in mode of preparation, quality, use, or effect.[5]

The problem was that nobody, least of all the clerks of the Internal Revenue Service, could make heads or tails of this specification. That is, nobody could tell what distinguished a patent medicine from a regular one. Congress attempted to single out drugs and medicines that were marketed directly to consumers, but in doing so they had little knowledge of the complexity of the trade they identified.

In taxing proprietary medicines in the same class as vice and beauty goods, Congress acted on a common knowledge distinction between legitimate and fraudulent medicines that had only recently taken shape. According to supposed categories legislators invoked, proprietary medicines were more like perfumes, alcohol, and cigarettes than something a sick person acquired to get well. But in practice, this category of suspect medicine was difficult to identify.

Wilson declared himself "worried beyond expression" over what had been done and what was coming next. After hearing the gruesome state of affairs, Lloyd requested to see for himself. He was led across the street to the regiments of lost clerks. Lloyd, posing as a person idly interested in medicine, wandered the rows, asking questions now and then about the work at hand. He found that Wilson was right, and that clerks were busy reversing each other's decisions as well as their own. With Wilson's permission, he took a sample of categorized labels back to his hotel room for study.

A few hours later, he returned and fulfilled Wilson's worst nightmare by declaring the situation hopeless. He figured the clerks had done their best given their knowledge of the industry and the know-it-when-you-see-it phraseology of Section 20—not simply all proprietary medicines but all medicines "put up in a style or manner similar to that of patent, trade-mark, or proprietary medicine *in general*"—and it went on from there. So it wasn't the clerks' fault, exactly. Nevertheless, they had thoroughly bungled the situation.

Lloyd advised Wilson to repeal all the rulings and get a statement from the attorney general clarifying the intent of the law, then base all subsequent rulings on that. The way he figured it, the only way any trumped-up classification would seem plausible was if the intent of the law had already been clearly established. As it currently stood, no one even understood exactly what the legislators had in mind. After a moment of dismay, Wilson entreated Lloyd to confer with the attorney general in specifying the intent of the law and the proper system of classification.

So it fell to John Uri Lloyd to devise the government's policy on taxing drugs and medicines to support the Spanish-American War. Lloyd went back to Cincinnati and spent two weeks devising tables of classification for the IRS and the attorney general, reluctantly distinguishing the so-called patent (proprietary) medicines from "true pharmaceutical preparations" and determining which of the former were taxable.[6]

The law required Lloyd to do considerable interpretive work in inferring and then justifying a series of categories he didn't believe withstood scrutiny. First, he tried to conjure the commonsense distinction between legitimate and spurious medicines that had informed the legislators, concluding that its specification of "patent" medicines was meant to target products sold directly to consumers from manufacturers. He explained that "patent" was a misnomer, since in fact the remedies were not patented (requiring disclosure in exchange for a limited monopoly on production) but kept secret—or at least, a full list of ingredients was rarely published on the label. In fact, the term "patent" derived from an earlier English usage in which a patent was an exclusive right to production granted by the crown rather than a limited monopoly for a new invention. So Bateman's Drops and Godfrey's Cordial, for example, were popular English patent medicines from which the Euro-American settlers derived their commercial tradition of retail medicines.

Lloyd suggested that the distinction between a patent (or proprietary) medicine and a true pharmaceutical preparation was better described as a distinction between (1) a medicine designed for self-treatment and (2) a pharmaceutical preparation, chemical, or drug for professional use. He suggested a descriptive term such as "domestic medicine," or "medicine used in self-treatment" should be used in lieu of the term "patent medicine." Medicines used for self-treatment were often secret mixtures, which used popular language and advertisements on the wrapper to appeal to consumers. Substances for professional use should list the contents in the vocabulary of a recognized scientific authority, including botanic origin, pharmaceutical composition, or chemical formula. They should use technical language, avoid popular names for ailments, and be labeled plainly, without advertising copy.

The supposed distinction between patent medicines and true pharmaceutical preparations represented the upshot of debates between physicians, pharmacists, and manufacturers of many stripes over patients' access to medicines. An influential group of physicians organized within the American Medical Association argued that doctors alone should prescribe medicines, which pharmacists would compound and dispense.[7] Reformers called these "ethical medicines" because they eschewed consumer advertising, adhered to industry standards established by the US Pharmacopoiea and the National Formulary, openly published all ingredients and formulas, and employed the specialized vocabulary of the physicians and pharmacists qualified to dispense them. Certain groups of manufacturers, such as Parke-Davis and Squibb, relished this arrangement. But it disadvantaged many other providers, including pharmacists who prescribed and prepared medicines independent of physicians, country stores stocking generic and proprietary medicines, and country doctors preparing their own remedies. Often these physicians also supplemented their supply with standard proprietary medicines, or packaged remedies that had already been compounded.

It was a testament to the success of reforming physicians that Congress asserted a distinction between legitimate and spurious medicines that seemed to hinge on access, targeting manufacturers who sold directly to consumers. But in practice, it was hard to distinguish between the two categories of medicine.

Lloyd did his best to make the imagined distinctions material. Adopting his best legalese, and working through many drafts with a surfeit of strikethroughs and insertions, Lloyd devised a many-point system to identify a proprietary medicine. First, he started with the materials themselves, attempting a system of identification based on packaging. Lining up many samples, he prepared a detailed description of each in reference to the others. These included: a "container (of wood) wrapped in plain paper [specimen] (A), a tin box unwrapped (specimen) (B), a bottle unwrapped by surrounded by a circular (specimen) (C), a (pasteboard) box unwrapped but surrounded by a circular [specimen] (D), a bottle wrapped by a pastel printed paper [specimen] (E), a carton of heavy paper [specimen] (F)." But after the first six, Lloyd concluded that these were among "hundreds of combinations" in general use.

Moreover, Lloyd concluded, the packaging of self-treatment and professional medicines was basically identical, so all distinctions had to hinge on the fine print of the label and accompanying copy. Lloyd compared labels and made copious notes, ultimately drafting a kind of police sketch meant to help clerks know a patent medicine when they saw it. In this expansive list of possible signs, titles with possessive case, trademark names, and claims of efficacy for a particular disease might all qualify a medicine as one marketed for self-treatment, and any one of these made it taxable. The best that can be said for Lloyd's careful classification is that it made as much sense as it could have given the porousness of the categories in question. Only the most litigiously inclined dared to challenge it when it was recorded as the official opinion of the attorney general one month later.[8]

In creating his taxonomy of taxable medicines, Lloyd inscribed a division between direct-marketed drugs and physician-prescribed ones that he found problematic. Lloyd knew what Congress didn't: reliable manufacturers of "ethical" drugs prescribed by regular physicians and dispensed by qualified pharmacists also marketed their products to physicians and pharmacists. After all, physicians, too, were consumers. If their salesmanship relied on alleged compliance with industry standards, it was salesmanship nevertheless.

Later, Lloyd published a pamphlet opposing the war tax that obviated his tortured distinctions between prescribed and consumer-marketed drugs. Too many classes of products were somewhere in the middle. Neither was "proprietorship in medicines" limited to those advertised to consumers. Rather, he argued, "some of the most typical patent and trade-marked medicines known are offered and sold only to the druggist and physician."[9]

Lloyd's Specifics were among the drugs that easily bridged the chasm between "legitimate pharmaceutical" and proprietary medicine. These whole plant drugs were dispensed by physicians and druggists, not sold directly to consumers. They were nevertheless wrapped in the Lloyd Brothers brand and advertised in trade journals. The classification Lloyd was obliged to devise made little room for his own products.

While reformers held that property in medicine was a corruption of consumer markets, Lloyd knew it to be fundamental to the business of physicians and druggists as well. He resented the suggestion that any class of medicines belonged in the same category as luxury or baneful goods. Rather, he argued, the true object of taxation should be any industry that enjoyed freedom from competition through monopoly rights. Here he anticipated (and resisted) a political economy of drug development that rewarded drug companies with actual patents. Lloyd believed manufacturers should distinguish themselves through the quality, pricing, and reputation of their products.

There were other assumptions of the stamp tax with which Lloyd might have taken issue. Taxing proprietary medicines as a luxury had the effect of penalizing people who resorted to self-treatment, especially those who lacked access to physicians. And physicians, too, might rely on retail medicines in their standard practice, especially if they employed a standard regimen of botanic preparations. Proprietary medicines were some of the best options available for people in rural areas. Lloyd's suggestion of "domestic medicine" as an alternative to "patent medicine" also specially targeted the home as a primary site of health care, and women as its primary providers. It is striking that one of the primary things that made a medicine taxable was its use of plain language on the label. It was a coup for reforming physicians, a repudiation of the botanic sects that had emphasized home health care, and an impediment for providers who had come to rely on packaged remedies. By comparison, Lloyd's critique or proprietorship in medicines was measured.

Why did the IRS call Lloyd? Lloyd wondered the same thing, probably partially out of concern that interference in the matter would jeopardize his business relationships. He suggested several other authorities, including eminent pharmacists, only to find Wilson had already consulted them. According to Wilson, they had declined to get involved, perhaps for fear of making enemies in the industry. Nevertheless, several committees of pharmacists and chemists from New York and Washington had already weighed in on the matter, leading to the morass of conflicting decisions in which Wilson found himself. So porous was the distinction between "proprietary" and "ethical" medicine that the leading experts in the field were as flummoxed as Congress and the IRS. Moreover, inasmuch as no one wanted to be taxed, and faced with a basically imaginary division between closely equivalent products, everyone would do things a bit differently.

In the eyes of other eminent pharmacists, some of whom had recommended him for the job, Lloyd was a reliable authority because of his training in chemistry, his experience as a compounding physician, and his laboratory orientation. Some of Lloyd's greatest contributions to the field of drug development were in the assays of alkaloids, the nitrogenous organic compounds of plant origin essential to the isolation of and production of plant drugs that remained primary, such as opium (morphine) and cinchona (quinine). At the same time, Lloyd's explicit association with Eclectic medicine and the Eclectic Medical Institute recommended rather than disqualified him to weigh in on the vagaries of sectarian and commercial botanic medicine.

Indeed, Lloyd defies the retroactive distinction between proprietary medicine makers who opposed regulation and so-called ethical pharmacists and physicians who supported it. Lloyd was a major advocate for the labeling of medicines, especially narcotic and addictive substances. He lobbied aggressively for the passage of the Pure Food and Drug Act, which was ultimately enacted in 1906. He insisted on the necessity of educational reform, including mandatory pharmacy school and laboratory instruction for apprentices. He served on the 1880 committee for the revision of the US Pharmacopoeia and was instrumental in the development of the National Formulary in 1888. He served as president of the American Pharmacy Association in 1887 and 1888 and published widely in industry periodicals. He also had the social stature to be a fishing buddy of President Grover Cleveland, for whatever that was worth.[10]

Ultimately, it was Lloyd's position at the margins of ethical and proprietary medicines that gave him authority. He was allied with botanic sectarians, not regular physicians, but this position qualified rather than discredited him among his pharmaceutical colleagues and government bureaucrats. He supported regulation and reform and had earned the esteem of the leading professional organizations. Yet he also advertised his medicines under the Lloyd Brothers brand. This practice rendered him guilty of proprietorship in medicine, even though his products were sold to doctors and pharmacists rather than consumers. Lloyd's business, like many others, belied the distinction between legitimate and spurious business the tax presupposed.

To add insult to injury, the IRS was asking Lloyd to justify a classification he had already rejected outright. During his service on the 1880 committee for the revision of the US Pharmacopoeia and again in the production of the National Formulary in 1888, Lloyd refused the suggestion that whole classes of medicines, like sweetened elixirs, or alcoholic preparations of medicinal herbs, be removed from the rolls. Rather, he believed inefficacy and abuse resulted from improper manufacture. He took the lead in developing 283 standardized elixir formulas. In his insistence on the standardization of botanic remedies containing alcohol, Lloyd refused the emerging division between ethical drugs and proprietary nostrums. He lost.

Ultimately, Lloyd was a reluctant participant in the bureaucratic and legal codification of congressional legislation that was itself a distillation of private reform agendas. The omission of his products from the categories the legislation implied presaged their disqualification by reformers. And in spite of his sense of his own contribution to history, the world as he knew it was passing away.

While Lloyd wrote copiously with an eye to preserving his memory, what the remnants of his archive communicate is the passing of an era, his business posthumously scrapped for parts. And so much of his legacy relates not to the records of his business but to his science fiction novel about journeying into the center of the earth through the mouth of a cave in Kentucky. Lloyd's hero, "I-Am-The-Man," becomes drawn to occult texts and ultimately disappears into a fantastical Underworld at the earth's core. After learning the secrets of matter and being, he returns as an ageless wanderer to the face of the earth, unknown and unknowable to the world he left behind.

Science fiction provided Lloyd safe ground to venture a critique of industrial society and its reliance on scientific practice. It also gave him a medium to explore connections and gaps between the complexity of nature and human knowledge of it. *Etidorhpa; or, The End of the Earth* has never gone out of print

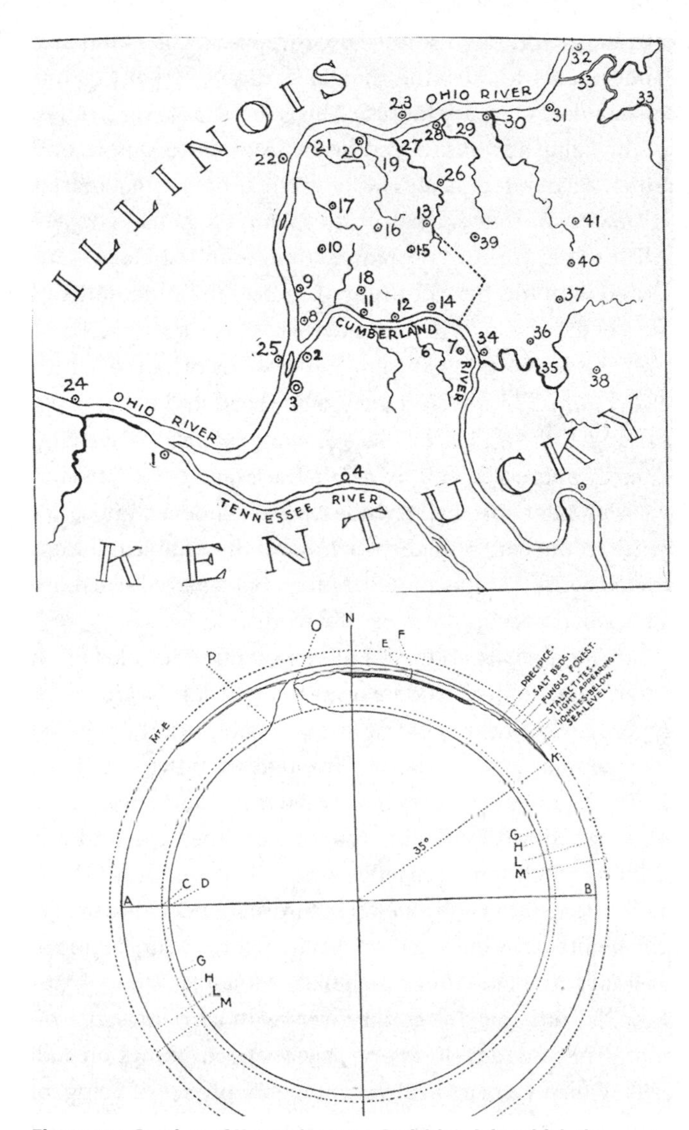

Figure 7.1. Section of Kentucky, near Smithland, in which the entrance to the Kentucky cavern is said to be located. Description of journey from K. [Kentucky] to P.—"The End of Earth," ***Etidorhpa,*** **John Uri Lloyd, illustrated by J. Augustus Knapp.**

since its publication in 1895.[11] Although some early reviewers accepted it as a true revelation of the occult world, Lloyd immediately disavowed any claims to veracity, declaring the book a literary fantasy. This is how most received it, placing it in the swashbuckling adventure tradition of Alexandre Dumas and Jules Verne. Other critics have noted its similarity to the utopian imaginaries of Edward Bellamy's *Looking Backward* (1888) and Étienne Cabet's *Voyage to Icaria* (1840). But unlike these works of utopian fiction, John Uri Lloyd's *Etidorhpa* is less concerned with the perfect order of society than the limits of human knowledge, and of the sciences in particular.[12]

According to the preface, John Uri Lloyd came into possession of the manuscript from the deceased Joseph Llewellyn Drury, who recounted the appearance of a mysterious stranger who appeared in his drawing room late one night. The stranger, who would be addressed only as "I-Am-The-Man," or, "I-Am-The-Man-Who-Did-It," insisted that Drury listen while he read aloud a manuscript documenting his voyage to the center of the earth—and then publish it after the passage of thirty years, when, he speculated, human beings would be more prepared to accept its revelations about the natural world.

As I-Am-The-Man tells it, he was abducted after violating the rules of an alchemical sect he had recently joined, endeavoring to publish its secret texts. After being prematurely aged by a chemical process, members of the sect led him on a circuitous journey along the Ohio River through the states of Illinois, Ohio, Virginia, and Kentucky and to the mouth of a cave near Smithland, where he proceeded with an otherworldly guide into the center of the earth and was forced to witness the limits of human knowledge.

Reflections on sensory perception and rules of nature follow. On the carriage ride through Kentucky before entering the cave, his first guide, who claims to be his son, answers aloud I-Am-The-Man's unspoken thoughts. When I-Am-The-Man wonders how it is possible, his son answers with a brief treatise on the limits of perception. The audacity of dreams, awareness of things unsaid, and insatiable curiosity of men were all evidence of higher phases of being, of which the sciences were a rudimentary beginning. Devotion to conceits such as biological age and human justice further compromise the validity of sensory perceptions. On his journey, I-Am-The-Man sees friends who cannot recognize him because he has been prematurely aged. They respond to his entreaties with confusion and derision, unable to understand physical change as resulting from anything other than of long passage of time. "It is as I predicted," avers his son. "You are lost to man." I-Am-The-Man's son is no more generous in his estimation of law, which he regards as based on hopelessly narrow judgments about what is possible: juries, lawyers, and judges are walking blind.

When I-Am-The-Man crosses the threshold of the cave, his journey turns from laws of man to laws of nature. At the mouth of the cave he meets his guide to the underworld, a creature without eyes covered in primordial slime. As I-Am-The-Man descends downward into the cave with the Eyeless Seer, he meets with sights that challenge his certainty about the rules of the natural world. Gravity ceases operation in perplexing ways, and he finds that he can jump great heights and fall as light as a feather. He finds a giant underground cavern filled with radiant light, although he saw his last ray of sunlight hours or days before. The Eyeless Seer explains that these phenomena do not contradict science so much as indicate its current scope.

Ironically, sensory perception, often regarded the very basis of scientific knowledge, cripples the seer. As I-Am-The-Man leads Drury through subterranean caverns of colossal crystals and walls of glass, he tries to explain that "in studying any branch of science men begin and end with an unknown." His dissertations on the laws of nature take him through the disciplines of astronomy, geology, mathematics, and chemistry, but I-Am-The-Man cannot comprehend the scope of the seer's insights because he is too baffled by his unprecedented sensory experiences. His empiricism is a handicap rather than a guide.

The Eyeless Seer has some thoughts about seeds. He tells I-Am-The-Man that a "grain of wheat is a food by virtue of the sunshine fixed within it." He tries to make his companion understand all animal and plant life as carriers of solar energy, and the farmer as a steward of photosynthesis. He predicts that when "inflexible climatic changes" have forced humans from the "bleak earth surface," they will find wild relatives of plants differently adapted to the climate, soil, and cultivation of the underworld.

Soon I-Am-The-Man finds a forest of colossal fungi with mushrooms that taste like strawberries. This is an aspect of the novel generally met with some mirth and supposition that John Uri Lloyd turned his pharmacological expertise to enjoyable ends. In fact, it was John's younger brother Curtis who was a mycologist with a tendency to wanderlust. While John was serving on professional committees and running the family business, Curtis was in Samoa, Jamaica, Italy, and Egypt on various collecting trips. In Lloyd's correspondence with the USDA Bureau of Plant Industry about *Helianthus radula*'s relation to *Echinacea angustifolia*, the government botanist offered to have one of his people get Curtis a mushroom with psychedelic properties called "Puff Ball" on an upcoming mission to the Yukon.[13] At any rate, John was generally inclined toward more modest visions. When I-Am-The-Man wonders at the mushrooms' gargantuan geometry, the Seer tells him the *Lobelia* seed of the upper earth is far more wondrous for being tiny.

Figure 7.2. "I was in a forest of colossal fungi. Handing me one of the halves, he spoke the single word, drink." ***Etidorhpa*****, John Uri Lloyd, illustrated by J. Augustus Knapp.**

Even so, I-Am-The-Man drinks from an inverted mushroom cap the size of a vat while the Eyeless Seer attends impassively, offering a survey of human fermented beverages from Tatar koomis to the "lascivious wine-bibbers of aristocratic society." Afterward, I-Am-The-Man meets a race of drunkards whose bodies have come to resemble their psyches, with elephantine limbs or facial features reflecting their errant desires. All warn him to turn back, but he continues with the Eyeless Seer to a precipice, where he plummets forward into the earth's inner space.

As he falls for minutes or hours or days, he stops breathing. His heart stops. And gradually he comes to rest before circling the depths and surging upward again to the surface of the earth, out of the cave mouth in Kentucky, and thence as an immortal wanderer in the Cincinnati area with designs on Joseph Llewellyn Drury's study.

Back above ground, I-Am-The-Man's debates with Drury lead the latter to question his conviction in the certitude of science. I-Am-The-Man relates horrible parables about the perils and atrocities of scientific habits of mind, including the career of a maniacal anatomist who turns to dissecting live babies and pregnant women after becoming dissatisfied with his ability to learn from the corpses he so casually exhumed from graveyards. Here Lloyd referred to a well-known black market in cadavers for medical education, often from the graveyards of working-class African American and Irish communities. In fact, opponents exiled the Eclectic Medical Institute from Worthington, Ohio, on the basis of such practices.[14]

I-Am-The-Man's scorn, and perhaps Lloyd's, was not reserved for perversions of medical education like the crazed anatomist's: what he reviled was the materialism of scientific investigators, which he considered indigent and destructive. In the materialism of modern science, discovery merely "transferred ignorance to other places." Further, he charged that science required a utilitarian logic that devalues individual lives, arguing that advances depend on the work of "investigators who overstep the bounds of established methods" such as the crazed anatomist.

In the process, humanity consumes itself, seeking immediate returns without any guiding values. A heartless thief, "science-thought" claims to aid humanity but destroys it by elevating the material, ultimately creating the conditions "to destroy the civilization it creates." The substitution of the term "science-thought" for "science" suggests that the broad-ranging practices he criticizes reflect less a series of practices or disciplines than a more fundamental orientation toward the world.

I-Am-The-Man compares scientific investigators to religious fanatics, adding: "Crazed leaders can infuse the minds of the people with their fallacies, and thus become leaders of crazed nations." Science, I-Am-The-Man insists, has become the new religion, its devotees "torturing, burning, maiming, and destroying humanity."

Drury rebels.

Intent on making Drury understand the perils of progress, I-Am-The-Man credits science with the most lauded and destructive artifacts of industrial society. "Who created the steam engine?" he inquires. "Who evolves improved machinery? Who creates improved artillery, and explosives? Scientific men. Accumulate the maimed and destroyed each year; add together the miseries and sorrows that result from the explosions, accidents, and catastrophes resulting from science improvements, and the dark ages scarcely offer a parallel. Add thereto the fearful destruction that follows a war among nations scientific, and it will be seen that the scientific enthusiast of the present has taken the place of the misguided fanatic of the past."

Figure 7.3. John Uri Lloyd reading *Etidorhpa*, undated photo. Lloyd Library and Museum, Cincinnati, Ohio.

In supplying "thought-food to fanatics," science enabled "the scald of super-heated steam," "explosions of nitro-glycerin," and the torture of animals, among other sins. Like religion, science laid "a road whitened with humanity's bones."

Drury recoils.

"We will discuss the subject no further," he replies. "It is not agreeable."

Etidorhpa was Lloyd's darkest image of the United States at the dawn of the twentieth century—an American West realized by railroads, fertilizers, fields, and slaughterhouses was figured as a Hades of industrial accidents, environmental toxicity, and animal torture. Yet his parable extended beyond a critique of industrial society to the systems of knowledge that enabled it.

I-Am-The-Man's subterranean voyage inverted Plato's allegory of the cave. In the latter, the philosopher formerly confined to see only the shadows cast on the wall is unshackled, initially blinded by light, but ultimately able to see objects as they are rather than as distorted by the movements of the sun.[15] In Lloyd's fantasy, the scientist learns by descending down into the cave, seeing light where there is no sun, and coming to understand the true operations of nature as independent of sensory perceptions and provisionally formulated laws of nature such as gravity and biological age. Rather than realizing that shadows are distortions of true objects, I-Am-The-Man crosses into a world where time and gravity operate differently, and men's bodies may materialize distortions produced in their minds. Rather than illuminating the principles of objectivity, Lloyd's man of science gains wisdom only by coming to terms with its blinkeredness and relativity. Discrediting sensory perception, legal institutions, and laws of nature, I-Am-The-Man's journey is an assault on the modern fact as the basis of objectivity and scientific inquiry, and of the consolidation of these practices in the social sciences as they were emerging in his lifetime.[16]

Lloyd's critique was moral and political as well as epistemological. With I-Am-The-Man as a mouthpiece, he questioned the utilitarian logic of "science-thought" that devalued individual lives in alleged service of a greater good. He advocated a keener awareness of the costs of progress. His emphasis, especially late in life, was toward values of conservation and stewardship.

But he was not the only apostle of preservation.

Roosevelt's ethic of conservation provides an interesting counterpoint to Lloyd's. For Roosevelt, the elk was a symbol of imperiled wilderness. "The open plains have already lost one of their great attractions," he lamented, "now that we no more see the long lines of elk trotting across them; and it will be a sad day when the lordly, antlered beasts are no longer found in the wild rocky glens and among the lonely woods of towering pines that cover the great western mountain chains."[17] He mourned the imminent extinction of the elk and the bison, but he did so in part because he believed sport hunting of the same animals preserved values of manliness such as bravery, risk taking, self-mastery, and violence. These, rather than the animals in question, were the objects of preservation.

Roosevelt's stories of the hunt were graphic and sensuous. He knew the smell of elk as something distinctive: strong, sweet, and pungent. Coming over a hill, he would smell the animals' bodies before he saw the tips of their antlers. The first sighting of the elk was as much about being seen as seeing, as the animal, walking through an open glade, suddenly spotted his unlikely predator poised some yards away. "Seeing us, but not making out what we were," Roosevelt wrote, the elk stepped boldly toward the group with a "stately, swinging gait . . . full of fierce and insolent excitement." The elk "stood motionless, facing us, barely fifty yards away, his handsome twelve-tined antlers tossed aloft, as he held his head with the lordly grace of his kind." Then Roosevelt shot the elk in the chest and raced forward as he turned, shooting him again in the flank. The second bullet was unnecessary, "for the first wound was mortal, and he fell before going fifty yards."[18]

Sometimes the act of killing was more drawn out. On a number of occasions he described how an elk would startle and freeze, then bolt as he rushed forward and fired on its flanks: or stop, sick and wounded, before he would break its neck with a fourth or fifth bullet. Even mortally wounded, an elk could run several miles. Thus a hunter with a repeating rifle, advised Teddy Roosevelt, should "run forward as fast as he can, and shoot again and again until the quarry drops."

Roosevelt's stories of the kill are followed by sensuous descriptions of the animal's body where it has fallen: landscape and body form a tableau of pristine wilderness. One fell in a glade at the edge of a cliff overlooking the "home of all homes for the elk": "a wilderness of mountains, the immense evergreen forest broken by park and glade, by meadow and pasture, by bare hillside and barren table-land." Another lay "among the young evergreens," his "huge shapely body . . . set on legs that were as strong as steel rods, and yet slender, clean, and smooth." Roosevelt studied the browns and yellows of the legs and body, the mane of long hair garnishing the neck and throat, and the symmetry of the great horns setting off the "fine, delicate lines of the noble head." On one of his prey he noticed a stab in his haunch, the wound from a losing battle that had cast him from the herd. He described dried mud clinging in patches to his flanks, evidence that he had been wallowing, "as elk are fond of doing." Then he cut off the head and bore it down to the train.

Ultimately Roosevelt got his own elk, *Cervus elaphus roosevelti*, or the Roosevelt elk, the largest of the four remaining North American elk subspecies in North America. And the values of "Nature, Youth, Manhood, and the State" he championed are inscribed in the rotunda of the American Museum of Natural History in New York, the museum he founded for the cause. Donna Haraway has emphasized the ways the values of masculinity and citizenship exhibited in the hunt and its trophies supported an imperialist, capitalist, white culture oriented toward bodily purity and the preservation of social order, reducing other objects to the status of prey. Education, science, and medical practice each shaped and conditioned these pursuits of natural order.[19]

Big game symbolized nature's power, and the nation's superior one. "Since writing the above I killed an elk near my ranch," Roosevelt wrote at the end of his chapter on elk hunting in *Hunting Trips of a Ranchman*, "probably the last of his race that will ever be found in our neighborhood."

John Uri Lloyd wrote differently about dead animals. In I-Am-The-Man's plain truth, a plate of veal was the "flesh of babies"; no life could be innocently conscripted for human pleasure. Lloyd also wrote lovingly of his boyhood dog, Turk, a career fighter who was sentenced to die for killing a neighboring farmer's sheep.[20] Lloyd, then twelve and just beginning his apprenticeship at Gordon's pharmacy, broke his own heart by choosing to shoot the dog himself rather than see it done. Like Roosevelt, Lloyd feared the progressive extinctions in the American West, intimating that species had symbolic importance to the nation that lay claim to them. But for Lloyd, animals were not so much exemplars of unspoiled nature or relics of the past as instruments of memory and rebukes of shortsightedness. As Krehbiel and Warkentin recalled the bones of bison strewn across the plains, Lloyd wrote of the caves of Kentucky as preserves of the big-boned mammals, the remains of which were progressively removed to museums such as those Roosevelt sponsored.[21]

Lloyd was a sensitive soul with an ecological sensibility, and an optimist who wrote a jeremiad under the cover of science fiction. But embedded in his novel was a redemptive vision of scientific practice oriented toward humility and self-awareness rather than progress and gain.

When Drury disbelieves I-Am-The-Man's story, he goes to consult his scientific friends regarding the plausibility of the traveler's claims as regards natural laws. One was Daniel Vaughan, a friend of John Uri Lloyd's whose obituary appears in the epilogue. Vaughan was a German immigrant and an esteemed chemist who died in poverty, having given himself over to investigations that would not pay. Vaughan concedes that I-Am-The-Man's tales do not accord with existing knowledge about the operations of gravity, but he is unable to discount the story just the same. When he pauses, Drury presses him as to his doubts. "I cannot find words in which to express myself," Vaughan replies, troubled. "I do not believe that forces, as we know them (imponderable bodies), are as modern physics defines them. I am tempted to say that, in my opinion, forces are disturbance expressions of a something with which we are not acquainted, and yet in which we are submerged and permeated." Vaughan was a cautionary tale but also Lloyd's friend, and the kind of scientific man he could admire; he never claimed to know more than he did, and indeed, struggled against the limits of what he could know.

"Beware of your own brain," I-Am-The-Man tells Drury, again and again.

What emerges as polemic in Lloyd's fiction is reflected much more dimly in his reckoning with the stamp tax and his reams of correspondence in search of a common weed: Lloyd considered science a style of knowing with limits too often obscured by fetishism in bureaucracy and law. In his inversion of Platonic truth toward radical relativity, Lloyd may have been ahead of his time. But then, anyone who tried to justify the tax code would have had to come to terms with the limits of reason.

In spite of the flurry of lawsuits that terrified Wilson, there was no great debate to be found in the case law for the stamp tax during the Spanish-American War. But historians should not go to every site seeking conflict. What this one reveals instead is bureaucratic irrationality in the making. The hall of clerks at the IRS was not so much a center of calculation or bureaucratic order as a locus of disorder and provisional arrangement. Rather than stabilizing meaning, the classifications Lloyd devised bring into question the verifiability and integrity of the material objects in which they trafficked. In this respect they had much in common with the classifications of plants, tools, and bones in the halls of the nascent Smithsonian Institution.

In retrospect, the hucksters and quacks were not so clearly identified. The expansion of markets requiring standardization of quality and value favored models of research science and regulation that ultimately attenuated both the biological and therapeutic diversity of plant medicines, recasting traditional remedies as foodstuffs, quack medicines, or weeds. Purging the pharmacopoeia of weeds was not quite as righteous as it would appear to the heirs of early twentieth-century regulation. Rather, it was a provisional solution to a lack of control—not simply over manufacturers, but over the plants themselves. And like all regulation, it was fundamentally exclusionary, targeting hucksters and quacks but reaching many other healers as well. The laboratory-research-based model of drug development that prevailed was not so much a logical or right outcome as a convenient solution to a problem of uncertainty.

Practices and materials that appeared marginalized nevertheless persisted without official sanction, either in pockets of local practice bypassed by global trade or in an alternative commerce of popular branded remedies. Commercial brands and advertisements functioned as alternative markers of authority, arrayed against chemical analysis and standardization. Meanwhile, forgetting traditional remedies enabled them to be rediscovered as the artifacts of twentieth-century pharmaceutical industry research, buttressed by US patent rights, however beholden to specific methodologies of extraction and synthesis. There is some evidence that, whatever the problems with variability, sourcing, and transit, plants are smarter than the vocabularies applied to describe and refine them. These possibilities interested Lloyd.

Neither I-Am-The-Man nor John Uri Lloyd hated science, but Lloyd, at least, came to hate the false certainty he believed it wrought. He believed the practice of science required perpetual transgression of boundaries and attention to the diversity and complexity of life, and these were the values of stewardship he advocated in his turn toward history and science fiction. In a political moment prioritizing certainty and might, Lloyd advocated uncertainty and restraint: an indication that he learned from the robust instability of the plants he spent his life studying.

In his baroque classifications of medicines for the Internal Revenue Service, his attempts to acquire a sufficient supply of roots from the vanishing prairie, and in the contours of his strange vision of the earth's core glimpsed through the mouth of a cave in Kentucky, Lloyd cast himself as a man of science who came to question what science could offer. Perhaps Lloyd was prophetic in imagining that humans would have to find alternative sources of food when the earth's surface became uninhabitable. I-Am-The-Man's conception of food as nothing more than metabolized solar energy too seems unusually prescient. But none of that was as important to Lloyd as what he didn't know yet.

Lloyd proposed to destabilize human knowledge systems, in part by questioning the temporal conventions through which they are organized: dusk and dawn, biological age, civilization and progress. When I-Am-The-Man's friends fail to recognize him because of his aged body, he realizes the limits of change as a concept. In human perception, change only occurs through time, and time draws its logic from biological processes. When he eats the mushrooms and stops the clock, he must regard as arbitrary the meanings he has applied to almost everything. With biology severed from time, he comes to regard the latter as a conceit to organize life in social terms.

The inscription on the verso of his manuscript, which he leaves with Drury, reads:

> There are more things in Heaven and Earth, Horatio,
> Than are dreamt of in your philosophy.

8 : WRITING ON THE SEED

Success is constant ongoing and ceaseless growth.
Life is motion forward. Motion is unavoidable in success.
— H. Ford. Syria 30-8-28 N. Makarem[1]

If there was a Neolithic revolution, it is still going on.
— Fernand Braudel[2]

The National Museum of Damascus preserves the world's first alphabet on a fourteenth-century BC clay tablet at Ugarit, the phonetic script a translation of an unwieldy accumulation of languages including Sumerian, Acadian, Hurrian, Cypriote, Aegean, and Hittite. In 1.3 by 5.1 centimeters, hundreds of cuneiform syllables became twenty-nine letters representing sounds. A less vaunted item in the museum's collection is even smaller, a wheat seed perhaps five millimeters in length. And it also preserves an alphabet, a poem of 113 words engraved on the hull by Sheik Nassib Makarem. The Druze carpenter and calligrapher, working in Syria (now Lebanon) during the Arab nationalist revolts of the early twentieth century, compressed poetry onto grains of rice and wheat, which he bestowed on kings, presidents, and once, Henry Ford: engraved with illustrations of the first three Model Ts and several of Ford's aphorisms celebrating forward movement as success (fig. 8.1).[3]

The calligrapher amused and delighted his public by simultaneously compressing and magnifying humanity on the head of a pin, producing the wonder that history on a grand scale could be written at the width of a hair. But his marvels, like the Ugarit alphabet, remind us of something more: that the seed itself is a marvel and has a grammar. In his early efforts to use the microscope as an aid to observation, Robert Hooke illustrated honeycomb-like structures of seeds he dubbed cells.[4] Two hundred years later, Ohio Secretary of Agriculture John Hancock Klippart subjected a glume of wheat in bloom to the same method, diagramming the anther in the process of extrusion (see fig. 8.2).[5] But the grammar of the seed is not merely morphological, nor expressed in the plant's life cycle. It also compresses millennia of evolution and the human en-

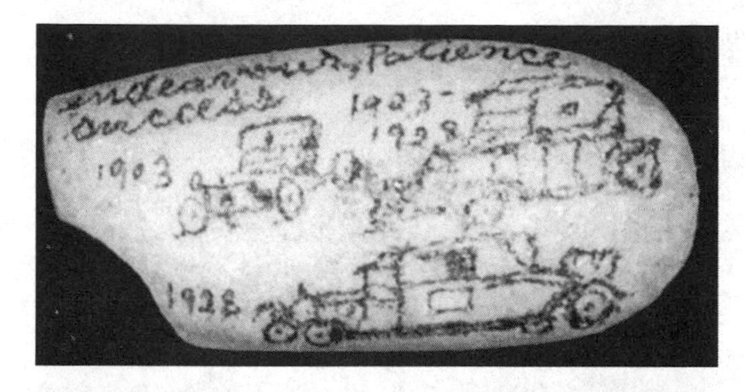

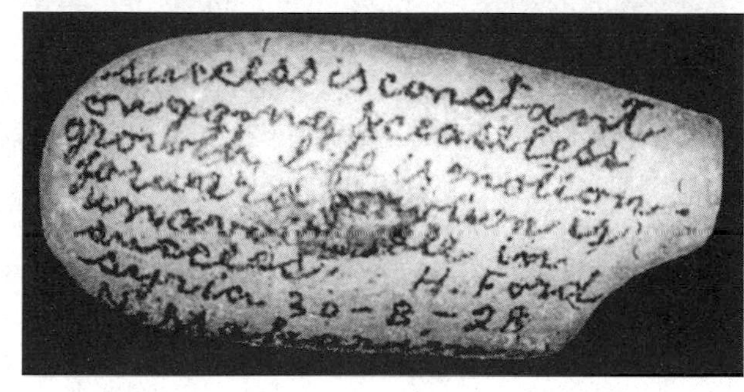

Figure 8.1. Grain of rice illustrated with (*A*) Henry Ford Model Ts, and (*B*) aphorisms, "'Success is constant ongoing and ceaseless growth. Life is motion forward. Motion is unavoidable in success. H. Ford.' Syria 30-8-28, N. Makarem." Used by permission of Samir Makarem and Sahar Kadi Makarem.

deavor required to support it. The wheat seed is a deep time technology, a wild cross of three grasses domesticated some 10,000 years ago along the banks of the Tigris and Euphrates Rivers and improved by successive generations of farmers ever since. So sophisticated it appears quotidian, the seed has been preserved through time to sustain human existence. Yet taken as a single unit, a grain of wheat is so common it has no value.

We tend to regard wheat as the Ur commodity because of its functional and physical characteristics. Functionally, it feeds the world. Human beings survive on a handful of cereal crops. Maize, rice, and wheat make up about half of the world's caloric intake. The United States is the largest exporter of two of these crops: corn and wheat.[6] Physically, wheat is easily exchanged. Reduced to its essential parts, it is small, dry, portable, and durable. It is homogenous and predictable, enabling trade in the absence of the product itself. Although there are different types and qualities of wheat, grading systems reduce these to standard values. These measures in turn facilitate the development of futures markets in grain, and the circulation of finance capital divorced from the material it represents.[7] In short, wheat earns an A grade for commensurability, the standard by which commodities are judged to be more or less perfect.[8]

But like all financial reductions, this one is fictional, obscuring the labor and knowledge required to produce the seed, and the ways in which the material itself can thwart production and exchange by behaving without regard for human needs. The modern construction of seeds as objects of research and development, institutionalized in public and private laboratories and subject to the disciplines of biology and genetics, obscures the long history of agrarian knowledge that undergirds them. Seeds are twenty-first-century genetic modifications, but also ancient technologies preserved and improved through generations of selection and improvement.

Instruments used to preserve and record plant matter, including herbaria, breed names, and genetic sequences create different grammars for seeds through varied modes of parsing, labeling, and representation. Each reflects different assumptions about natural and human creativity/agency and the proper political economy for managing material and intellectual resources. These assumptions are in turn supported by varied conceptions of time, alternately fixed to imperatives of ecology, civilization, harvest, and sale. Through the preservation, documentation, and improvement of seeds, their custodians regenerate these temporalities, forging new histories and narratives of development. Seeds may figure as breakthrough technologies, heirlooms, or our daily bread.

What would it mean to reconceptualize a seed as a deep time technology? Approaching a grain of wheat in deep time provides a way to explore the fluid, collaborative, and multigenerational aspects of agricultural improvement; its ephemeral, fugitive, and lost-in-translation qualities; and the stubbornness of the seed itself, which operates according to its own rules for when to live and die. The cyclical and repetitive qualities of cultivation and regeneration bring into question concepts of biological innovation supposing linearity, progress, and property, and the progressive histories that support them.

Perhaps because historians peddle time for a living, they are often reluctant to question its value. Historians use time to order narrative, to organize experience, and to expose ideas as contingent, bound by the milieus in which they occur. But time's fixity is too often taken for granted. Inasmuch as it imposes structure and explanation, time is the *problem* rather the *given* of analysis.

In the past decade, anxieties about global climate change have inspired historians to consider human social and political activity with more expansive chronologies and categories. Perhaps the most prominent is that of the Anthropocene, first suggested by atmospheric chemist Paul Crutzen and biologist Eugene Stoermer to delineate a new geologic age characterized by human impacts.[9]

The periodization of the proposed Anthropocene remains a matter of some debate.[10] According to Crutzen, it is coterminous with the advent of industrialization: James Watt's steam engine in 1784, the burning of fossil fuels, and concentrations of carbon dioxide and methane revealed through stratigraphy and associated modes of analysis. Others propose the advent of global capitalism as a more plausible causal event precipitating a turn to new sources of energy. A minority of theorists provide a longer trajectory, identifying the advent of agriculture some 10,000 years ago as the tipping point of human agency. The systematic cultivation of plants in turn provoked extractive and energy-intensive industry exploiting wood, coal, petroleum, and gas. More typically, geologists identify this period as the transition from Pleistocene to Holocene, when the gradual warming of the globe in the wake of the Ice Age enabled domestication of plants for human consumption. Yet if climatic changes beyond human control enabled plant domestication, the stabilization of temperatures nevertheless precipitated a fundamentally destructive agricultural-industrial way of life. It's a bleak proposition: either industrial capitalism is destroying the planet or human beings are an invasion species. In either case, these periodizations echo an old history of unintended consequences, the magnitude of which we've only recently come to realize.

Arguably the challenge of suprahuman time scales has been greatest for historians of modern world systems. In the face of extinction, categories of globalization and empire lack explanatory power, and capitalism figures less as an engine of inequality than as a system without limits. As Dipesh Chakrabarty has supposed, perhaps personal agency (freedom) and geologic agency were coterminous developments: human beings became free to extinguish themselves. By implication, we must question paradigms of human power, justice, and liberation framed in relation to empire, modernity, and the separation of natural and human history. We must "mix together the immiscible chronologies of capital and species history."[11]

On examination, however, chronologies of capital and species are not as immiscible as they may seem. Scientific timescales first and foremost are artifacts of human investigation, inevitably inflected by human needs and desires. As debates over the periodization of the Anthropocene suggest, even attempts to unseat humans as heroes of history may be quite reflexive, and, indeed, anthropocentric. Far from disrupting prevailing notions of stratigraphy or chronology, the designation of a new geologic age affirms their stability, and indeed justifies technological interventions to correct human impacts. That is, the Anthropocene is above all a call to action, and a new point on the timeline rather than a rejection of the timeline itself.

Nineteenth-century geologic timescales too were products of scientific engagement with political economy, prioritizing extraction and exploitation rather than sustainability. In the 1820s and '30s, the deliberate and incidental unearthing of human and animal fossils, cave drawings, and artifacts galvanized attempts to date the long course of life on earth. In fact, the Anthropocene is an addendum to a nineteenth-century geologic timescale devised in reference to archaeological and fossil remains. In multiple editions of *Principles of Geology* (1830–73), Charles Lyell used fossil mollusk assemblages in western Europe to reperiodize the earth's settlement, building on late eighteenth-century divisions of the earth's crust into Primary, Secondary, Tertiary, and Quaternary layers. Lyell subdivided the Tertiary and coined the terms Pleistocene and Holocene to refer to epochs within the Quaternary.[12] Meanwhile, Danish archaeologist Christian Thomsen devised a chronology of human ages based on layers of excavated materials and hand tools: stone, iron, and bronze.[13] In the study of biology, Charles Darwin's *Origin of Species* (1859) inspired new linkages between systematics and a theory of evolution, turning inward from morphology to rules of functional and physiological development. The combined effect of these movements was to provide separate, if overlapping, chronologies of life in the fields of geology, archaeology, and biology.

Appended to the Pleistocene and Holocene, the Anthropocene regrafts geologic time onto a schema with materialist underpinnings, leaving the latter intact. Among the most powerful of such narratives of development is the archaeologist V. Gordon Childe's "Neolithic Revolution," which attributed all civilization to the invention of agriculture in the ancient Near East. Childe began using the term "Neolithic" in the 1920s, featuring it in his 1936 autobiography *Man Makes Himself*. Childe's teleological bent echoed those of Karl Marx and the American ethnologist Lewis Henry Morgan. Morgan posited that cultures evolved from barbarism to civilization through technological development originating in plant and animal domestication.[14] These heroic accounts of human development privileged moments of rupture and revolution, reading Euro-American civilization as the apex of all history to date. In these stories of progress, the question was how and why things changed, not how and why they stayed the same.

In recent decades, archaeobotanists and paleoethnobotanists have given us new tools for thinking about continuity and gradualism.[15] A post-Pleistocene global temperature increase allowed the expansion of wild grasses, facilitating a shift from hunter-gathering to complex foraging and gradual plant cultivation. And although we know wheat is a wild cross of three grasses, collected by human beings and reproduced 10,000–12,000 years ago in the Near East, recent research suggests simultaneous domestication of plants in Central America and southern China. Natural selection, speciation, and adaptation enabled practices of artificial selection and hybridization. Common bread wheat originated from a cross of cultivated tetraploid emmer wheat (*Triticum dicoccoides*) and diploid goat grass (*Aegilops tauschii*) approximately 8,000 years ago.[16] Through slow migration, drift, natural selection, and deliberate improvement, wheat evolved into highly heterogeneous landraces improved by successive generations of farmers.

These studies of gradualism and continuity suggest a history in which humans are not prime movers, and many humanists have taken these insights to heart. Historians inspired by paleobiology have called for more contextualized investigations of human agency, exhorting historians to break the barrier of sacred time and investigate humanity in light of insights from neuroscience and species biology.[17] Political theorists have pursued new metaphors of temporality, premised on the notion that we can learn better behaviors from nonhuman actors. Gilles Deleuze and Félix Guattari imagined culture as a rhizome, nomadic and profligate, opposing it to a root-tree structure of linear chronology and causation.[18] William Connolly has imagined a world constituted of multiple force fields or "tiers of becoming," in which the frictions of geologic time, climatic time, and capitalist time generate uncertainty and disequilibrium.[19] Attending to nonhuman agency and other forms of cross-species thinking, these theorists deploy images of connectivity, entanglement, and flux against teleological narratives of progress and growth.

These theorizations complement histories of time as a cultural artifact. Peter Galison has argued that Einstein's studies of time were inspired by the problem of synchronizing railway clocks.[20] If indeed the theory of relativity emerged from military-industrial imperatives of maritime and rail travel, it should not surprise us that most studies of temporality in the twentieth century have revolved around the clock as an instrument of discipline. E. P. Thompson's essay "Time, Work-Discipline, and Industrial Capitalism" interpreted timekeeping as an imposed habit of compliance to industrial norms of production.[21] His analysis provoked a flood of sociological literature on shop floor management, often focused on Frederick Winslow Taylor's innovations to speed up production.[22] Cultural critiques of capital remade Taylor's stopwatch, and Henry Ford's assembly line, as symbols of dehumanization, not the "constant ongoing and ceaseless growth" for which Makarem honored Ford. In Fritz Lang's *Metropolis*, the worker-hero is crucified on the heavy hands of a clock, which he must wrest backward for the entirety of an overlong shift: "Father—! Father—! Will ten hours n e v e r end——??!!" Time stretches the hero, and his transcribed utterance, on a rack. In Charlie Chaplin's *Modern Times*, the Little Tramp's distraction twists the assembly line into a manic spiral as each worker rushes backward along the conveyor belt to correct his missed piece. Ultimately, the unlucky tramp is sucked into the machinery itself and turned over the cogs on the wheel. In sum, many historians of capitalism have insisted that industrialization changed time, making it the apparatus of work-discipline. Ford thought life was motion forward, but his workers' lived experiences of acceleration and disruption belied his claims to linearity and progress.

Other historians have attempted a more global analysis of capitalism's time than individual histories of labor allow. Before the theorization of capitalism as a world system, Fernand Braudel contrived to narrate the whole history of the quotidian world, making the *longue durée* a canvas for nature's persistence and power in human history.[23] More recently, advocates of commodity chains analysis have constructed spatial narratives that render temporality subsidiary to the logic of capital flows, with objects moving through a predestined life cycle from production to consumption.[24] If these representations naturalize and reproduce capitalism's time in lieu of explaining it, in this they are hardly unique.

Rather than reflecting time as it is, social and political knowledges produce models of temporality. Historians of concepts have been particularly attuned to the ideological characteristics of these models.[25] Perhaps the most explicitly attuned to temporality was Reinhart Koselleck, who suggested a "saddle time" of "threshold time" between 1750 and 1850, marking a shift from sacred to secular prognosis, and by extension, from universal to world history. As sovereign states wrested powers of prediction from the church, they oriented their authority toward political calculation rather than eventual apocalypse, providing a scale of possibility that allowed for foreseeable outcomes. Nevertheless, Enlightenment expectations of human liberation placed the future always out of reach: a "perennial imperative [*sollen*]," a "finite not-yet." Thus the progress to which they aspire is necessarily ambivalent.[26]

History as a discipline emerged from this mixture of "rational prediction and salvational expectation," necessarily subservient to the realization of political ends. Its chronologies consisted not so much of facts as possibilities and prospects plotted in time: "past conceptions of the future," "futures past." We organize time by tagging experience: as continuity, repetition, reiteration, change, crisis, revolution, progress, period, era, and epoch. These chronologies prioritize accounts of secular progress that assert human creativity and agency in structuring causation. Often they draw support from the varied temporalities crafted to explain life in geologic, archaeological, and biological terms. Bound to temporalities of crisis and revolution, progressive logic bleeds across many domains of knowledge.

But multiple ways of thinking about time persist within progressive histories: thoughts, deeds, plans, events, nature, mortality, politics, and institutions inhabit different registers of possibility. Attending to this bricolage can help us find new ways of thinking with time. Ultimately, the greatest challenge of nonhuman time scales is not their suggestion of human geologic agency or probable extinction, but their invitation to dismantle the frames we've constructed to make large-scale movement legible.

The domestication, preservation, and improvement of seeds trouble conventional timescales. Naturally self-replicating organisms, seeds nevertheless require human stewardship and intervention to make viable food supplies. Plant domestication is the purposeful selection and preservation of seeds: a technological practice with a 10,000-year vintage. In the field, farmers preserve seeds in situ, regenerating them from year to year and maintaining their diversity through varied modes of cultivation and exchange. This deep history of agrarian knowledge is the foundation of all subsequent interventions to preserve and improve seed.

Systematized collection of seeds for study also has a long and varied history. In the European context, one of the most durable forms of documentation and preservation is the herbarium: a systematic collection of dried plants, generally assembled in a book, box, or cabinet. Herbaria originated in traveler's accounts and devotional books and were repurposed and expanded for natural history study in sixteenth-century Europe. Like the herbals of earlier centuries, herbaria draw on reclaimed Greek and Latin texts on medicinal plants, as well as contemporary field specimens, but they differ in observing, cataloging, and describing nature for its own sake. Gradually, herbaria evolved from memory aids to tools of study and centers of documentation.[27] These organized and expanded sites preserved the labor of collecting linked to European maritime exploration, colonization, and empire.

Although format and technology are largely static, the content of herbaria have changed over time, principally in the modes of description and identification of plants. Early herbaria tended toward simple collection, but as the discipline of natural history developed in the sixteenth century, collectors added thick descriptions of environment to the specimen pages. A century later, these contextual renderings gave way to an exclusive focus on plant morphology, linked to Linnaean practices of classification and systematics in the eighteenth century. Modern projects of taxonomy superseded folk taxa, which identified plants in local context. If ethnographic and environmental data were lost in the process, this was a by-product of an effort to come to terms with the global scale of plant life by crafting a coherent universal system to describe it.[28]

In some respects, the modern gene bank is the logical extension of Linnaean aspirations to universal order represented by the herbarium, perhaps even more so than the great European botanic gardens, which turned their efforts to the purposeful transplantation and acclimatization of diverse plants.[29] The herbarium and gene bank attempt a textual and material library of world biodiversity, translating geography into an ordered list of taxa and numeric identifiers.

One realization of this universalizing system is the Linnaean binomial, retained into the twenty-first century as the lingua franca of plant scientists, and indeed their only common language. (In collecting expeditions staffed by Tajik, Kazakh, Armenian, Georgian, Russian, Kiwi, Greek, and Syrian scientists, we generally bellowed across the field in Latin, condensing a welter of information into a single identifier: *Aegilops crassa! Aegilops tauschii! Hordeum brevisibulatum!*) Bagged, threshed, cleaned, and lodged in a gene bank, a plant's local identity becomes subsidiary to the binomial and accession number in the database.

The persistence of the herbarium as a working technology is another indication of continuity between eighteenth- and twenty-first-century sciences of preservation. As Brian Ogilvie notes, a sixteenth-century botanist would easily recognize and use a twenty-first-century herbarium, and the reverse is also true. One "reads" the herbarium in a remarkably consistent way across time.[30]

The herbarium of the International Center for Agricultural Research in the Dry Areas (ICARDA) in Tel Hadya, Syria, provides a case in point. Until 2011, the herbarium consisted of a fifteen-square-foot room adjacent to the 150,000 seed specimens preserved in its gene bank: a series of temperature-controlled vaults with rows of cartons and packets containing live seed. Considered part of the Genetic Resources Section (GRS), the herbarium's focus is crop wild relatives (CWR) in the ICARDA mandate region, extending west from Mauritania to Morocco, eastward across the Mediterranean basin, through eastern Europe, into the Middle East, West Asia, Central Asia, and the Caucasus. The largest numbers of herbarium specimens are from Syria, with many from neighboring Lebanon, Jordan, Turkey, Iraq, and Iran: the countries of the so-called Fertile Crescent. Since 1977, collectors have gathered some 16,000 herbaria samples in seed collection expeditions, during regional vegetation surveys, on informal visits to ancient sites such as Saint Simeon and Palmyra, and

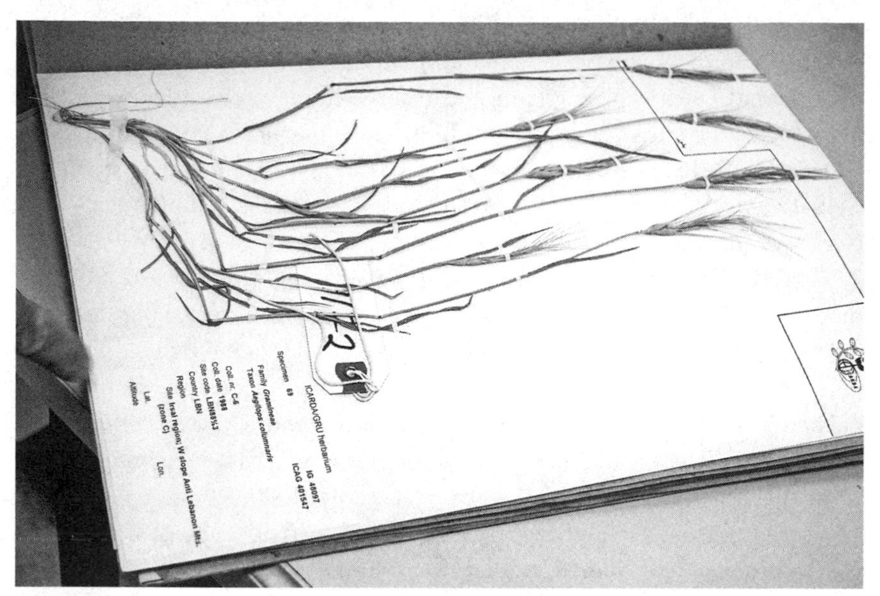

Figure 8.2. Type specimen of *Aegilops columnaris* in the ICARDA herbarium, Genetic Resource Unit, Tal Hadya, Syria, July 2010. Photo by Courtney Fullilove.

through field multiplication and regeneration at Tel Hadya. The herbarium includes wild relatives of both cultivated and forage cereal and legumes.[31]

Old technologies may serve new masters. Within the past twenty years, plant geneticists have turned their attention to herbaria such as ICARDA's as sources of genetic material. Rather than simple study collections, herbaria may contain genes for resistance to biotic and abiotic stresses, including pests and climate change. In recent decades herbaria have also been plumbed as the material for "gap analysis" of regions where few seed samples have been collected.[32] Nevertheless, funding for herbaria has lagged behind that of gene banks, signaling the declining status of botany as a scientific discipline, and of the herbarium as a technology of preservation and documentation. At the Vavilov Research Institute, the herbarium had been disaggregated from the seed collections sometime in the 1960s, with its organizational structure lost in the mix. Boxes upon boxes of uncataloged pages lined the shelves, with no manpower or funding to sort them out again.

Gene banks exhibit a different relation to the imperatives of time and capital. Generally separated into short-, medium-, and long-term storage, gene banks arrest material in a live state, ensuring its viability for future use. In their commitment to deny natural processes of generation and decay, gene banks defy time, insisting on a coterminous past, present, and future.

Often these claims to timelessness cluster with ones of heritage, rarity, and the deep past, with prominent international gene banks such as the Svalbard Global Seed Vault pledging to preserve humanity's agricultural heritage against the encroachments of conflict and ecological ruin. Established in 1984 under the leadership of conservationist Cary Fowler and managed by the Global Crop Diversity Trust, the facility on the remote island of Spitsbergen, Norway, is meant to serve as a primary record of global biodiversity.

These initiatives cross a casual strain of end times thinking with the high modernist conviction that scientists have figured out a solution by preserving a copy of the world's plant genetic heritage under one hundred meters of permafrost at the North Pole. Yet however striking the spectacle of a trapezoidal hatch shuttering a vault of seeds in the arctic tundra, these claims for the deep historical legacy of seeds are not new, nor waged with total innocence.

In the nineteenth-century United States, the dream of agricultural improvement via genetic archaeology entailed a schematic variety of historical geography, well represented by recurring submissions of salvage wheat to the Agriculture Department. The search for ancestral seeds invoked the time of civilizations, situating Europe and the United States at the apex of development. Egypt was a popular target of fantasy. Egyptians, the story went, had carried the art of agriculture to a state of perfection, sustaining their own dense population and making Egypt "the granary of the world." The grains of wheat found in mummies would see the United States realize the same destiny.

Even for those who resisted these fantasies, wheat was rarely just a staple crop. In his 1859 treatise, Ohio Secretary of Agriculture John Hancock Klippart called it the "true and unequivocal symbol of civilization, and consequent enlightenment and refinement." Wheat induced "man to forget his savagism, abandon his nomadic life, invent and cultivate peaceful arts."[33] In Klippart's analysis, American agricultural development took on world-historical significance. Ironically, hopes of American agricultural improvement rested on a narrative of decline. Whereas modern-day Egypt was an object lesson in soil exhaustion and a nightmare of America's agricultural future, as the cradle of civilization it represented a myth of agricultural prosperity and innovation.

When Klippart produced his synthesis on the wheat plant, its place of origin was as yet unknown. He studied Alexander von Humboldt's theories, positing the origin of cultivation at the place of the greatest number of indigenous known species of the same genus. By this reckoning, Persia and India were probable places of origin. Travelers' accounts and submissions of primitive wheats encouraged further study and speculation, while experimental trials with *Aegilops* encouraged further work on the mechanisms of domestication.

Figure 8.3. Map showing James Henry Breasted's identification of the Fertile Crescent in Breasted and Robinson's *Outlines of European History, Part I: Earliest Man of the Orient, Greece, and Rome* (Boston: Ginn, 1914).

Later the supposed place of origin for the cultivation of wheat would come to be called the Fertile Crescent, the accordingly shaped area of fertile land extending from the eastern Mediterranean through the valley of the Tigris and Euphrates. Chicago Egyptologist and philologist James Henry Breasted improvised the term "Fertile Crescent" for a world history textbook in 1914, entering the term during the typesetting phase of the textbook's publication.[34] For such a fleeting identification, it has proved remarkably hard to dislodge from the public imagination. While agriculture originated there some 10,000 years ago, archaeological evidence intimates gathering of wild emmer wheat 19,000 years ago, suggesting that humans collected wild grains for some 10,000 years prior to domestication. Common bread wheat, or hexaploid wheat, probably emerged as a hybridization of emmer wheat (*Triticum turgidum*) and goat grass (*Aegilops tauschii*) in southeastern Turkey or northern Syria around 8,800 to 8,400 years ago.[35]

Breasted's Fertile Crescent was an accessible handle, but it also implied a certain political geography. Like any textbook, Breasted's *Outlines of European*

History reflected broad assumptions about history and politics. The most obvious of these is that the legacy of the ancient world, regardless of its geographic indicators, belonged to Europe. James Harvey Robinson, Breasted's collaborator, was a proponent of the "New History," which contended that positivist historians had much to learn from the newer sciences, and especially evolutionary theory. He believed that the study of history should be a force for social change. While Robinson's political sympathies, including their social Darwinist cast, were of a piece with those of other Progressives, Breasted had an additional agenda, at least implicitly. Penned on the eve of World War I, Breasted's celebration of the agricultural wealth of ancient empires reflected a fear of civilizational decline and a latent sympathy for imperial control of natural resources.[36]

In other words, the narrative of civilizational advance also implied one of decline, justifying European and American imperial ambitions. "In countries where the agricultural art, or rather the culture of the wheat plant, has fallen into disuse," wrote Klippart, "there has civilization also retrograded; and were it not for commerce with enlightened and refined nations, several countries would speedily relapse into all horrors of absolute barbarism." Without agriculture, there would be no industry: no iron smelting for plows, no machines for harvest, no transportation to market.[37] In societies without wheat, there was a total absence of human creativity and innovation.

However beloved this creation story of agriculture, contemporary research rejects the notion that agriculture appeared spontaneously in a single region. The changeability of our stories of origin should suggest revision of our concepts of innovation, but the shibboleth of unilateral progress is a powerful one. Logics of national greatness and ancient patrimony remain prominent in collecting expeditions, which mingle historical fantasy with bioarchaeological data. Ideologically, fantasies of salvage wheat reflected a belief that Americans were the rightful heirs of ancient civilization. Like Childe's and Breasted's narratives of domestication, these relied on evolutionary narratives linking agricultural improvement to civilizational development.

In nineteenth-century America, a casual conflation of national and civilizational time justified the massive importation of seeds and cuttings for the benefit of American farmers, stoking the fantasy that the United States could become the granary of the world. It was a fantasy the nation ultimately achieved, at some cost.

The secular time of progress depicted by Klippart supported the more fine-grained chronology of property forged by entrepreneurs and breeders; it made property and innovation civilized. In the later nineteenth century, the USDA partnered with farmers and seed companies to produce new and improved varieties using single line selection and hybridization. Only in the 1920s did corporations begin to exercise a preference for privatized breeding using closed lines and intellectual property protection. Double-cross hybridization escalated the commercialization of seeds by producing a single season of "hybrid vigor" followed by a gradually decaying seed stock.[38] These were seeds with expiration dates, in need of constant refurbishment, an obsolescence that accrued to the benefit of those who sold them. The proliferation of varietal names such as those surveyed by Carleton, Clark, and Ball marked the truncated temporality of modern cultivars, amplified in the export of Norman Borlaug's high-yielding hybrids in the 1960s.[39] Lineage signaled continuity and improvement, but also obsolescence.

Although gene banks make claims to preserve the world's seeds, their primary function is to support public-sector breeding, and in certain respects they remain constrained by the political frame in which they were realized. As one of fifteen nonprofit organizations gathered under the Consultative Group on International Agricultural Research (CGIAR), ICARDA is the heir of Norman Borlaug's Rockefeller Foundation–funded research in the Mexican Agricultural Program, credited with spawning a "Green Revolution" in Asia. In addition to providing a record of world biodiversity, gene bank collections provide raw material for breeders who pursue improved varieties resistant to pests, drought, and disease. Dispersed in Lebanon, Morocco, and Norway, ICARDA's gene bank currently serves only one of its two functions: preservation, not circulation. In the absence of unified staff and resources, responding to researchers' requests for germplasm is impossible. Yet genetic resource specialists perceive the utilization of their collections to be as essential as preservation, motivating ICARDA to request the first ever withdrawal of seeds from Svalbard to populate its new facility in Rabat, Morocco.[40]

As seeds move from cold storage into circulation, their temporality shifts again. Preserved in long-term gene banks, seeds represent claims of geographic origin, biodiversity, and long-term stewardship by communities of farmers. As raw materials for improved varieties, they acquire new abbreviated timescales linked to moments of generation, distribution, reproduction, and sale.

Once imposed on the seed, these grammars of improvement become the invisible artifacts of progressive social and technological order. But we can make them visible again by focusing on the reorganization and occasional liquidation of specimens and the rules of exchange grafted onto them. Technologies of preservation are as fragile as they are persistent, as the foregoing history

Figure 8.4. Nawras El-Hajj in the ICARDA gene bank, Tel Hadya, Syria, July 2010. Photo by Courtney Fullilove.

of shells and tarpaulins, glass vented cases, storerooms, greenhouses, grain elevators, and field margins demonstrates.

As violence consumed Syria, ICARDA staff endeavored to arrange transport of duplicate specimens from the gene bank to facilities in Lebanon, Jordan, Tunisia, and Norway. The herbarium had a different fate. In May and June 2012, remaining Genetic Resources Section staff members at ICARDA packed the twenty-plus metal cabinets of the herbarium into a truck and directed their evacuation from Tel Hadya, through military checkpoints, to a safe house in Aleppo. Their current status is unknown.

Farmers in the field cannot call on even these modest protections, with their resources, livelihoods, and survival threatened by pervasive destruction. If those who remain become targets of reconstruction, the national and international agencies that said them will confront now familiar decisions about how to administer agricultural development in postconflict regions. In these spaces, an array of NGOs stake out agricultural futures linked to seeds, equipment, and chemicals of varied commercial and national provenance. The seeds these farmers cultivate reflect practices of naming and claiming devised in twentieth-century institutions of research and development, and the legacies of conflict that brought them into circulation. Awareness of these histories can support more sensitive political interventions.[41]

But this is primarily a book about how we can learn from nineteenth-century agriculturalists, who tended to the fields while the rocks and metals and plants around them acquired new temporalities linking settlement layer to cultivation to civilization, ultimately marking a space for something called "biological innovation." Within these modern calcifications of time around capital and country, we can find other ways of knowing about seeds that trouble their conscription as objects of property.

If we regard seeds as artifacts, we implicitly recognize them as products of human labor, not nature. Students of material culture have long since abandoned an emphasis on decorative arts in favor of more functionalist and cultural approaches, but ironically aesthetics may be useful in analyzing the most quotidian of objects: staple grains. If we regard them as objects of art, we acknowledge that the labor required to produce seeds is skilled, and that appreciating it requires us to cultivate critical and aesthetic sensibilities similar to those used to evaluate the workmanship of furniture, silver, or ceramics.[42]

In his treatise on the wheat plant, Klippart gave extensive attention to the conditions required for the impregnation and hybridization of plants. Deliberate and systematic hybridization was a well-established practice by the early nineteenth century, and experimenters recognized it as a path to additional knowledge about plant morphology and behavior. Klippart could summarize the leading research of Mr. Maund in Bromsgrove, Warwickshire (England), and Daniel J. Browne of the US Patent Office in Washington. Both gave detailed accounts of how to trick wheat out of its desire to self-pollinate, including the proper temperature (warm) and times of day (between ten and twelve o'clock). By manually holding the head of the female downward, carefully opening the glumes, and cutting off the anthers with very sharp pointed scissors (taking care that no anther is permitted to touch the pistil of the same head), pollen grains from the anthers of the male specimen could be immediately applied to the pistil of the glumes from which the anthers had been removed.[43] Maund offered his own experience to Browne in preventing self-fertilization in wet and hot weather, when much pollen is released in the chaff:

> Often in moist weather have I felt much interested, when, wanting pollen, I have held the straw and bottom of the ear in my warm hand for two or three minutes, watching for a crop of anthers. Quickly, the ripest of them, stimulated by the warmth, would peep out from their seclusion, and, gently rising, give me a chance of capturing them ere they scattered their contents over the expectants beneath them. Sometimes, on laving these excited ears, and returning to them after ten or fifteen minutes, I have found several

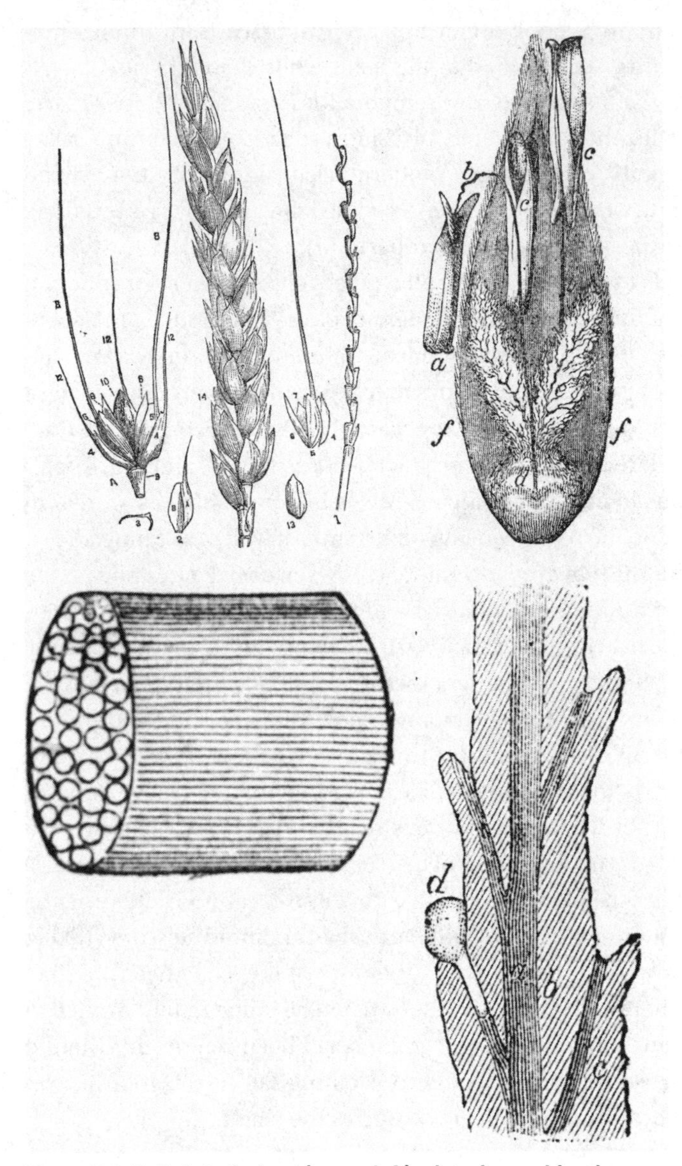

Figure 8.5. Spikelet of wheat (*upper left*), picturing rachis, glumes, and awns; magnified glume of wheat in bloom (*upper right*), picturing ruptured anther (*a*), filaments (*b*), anthers not yet extruded (*c*), ovarium (*d*), pistil (*e*), and glume (*f*); magnified and divided portion of anther (*lower left*), showing arrangement of pollen grains, which when mature sheds pollen on the pistil (*lower right*). From John Hancock Klippart, *The Wheat Plant* (1859).

> anther-cases as empty as balloons, dancing to the breeze, as if joyous that in my absence they had scattered every pearl they possessed.[44]

Maund offered a different sort of poetry than micrographs or base sequences of DNA, giving a detailed and tactile account of the wheat plant's anatomy. The empty anther cases have grace and verve, happily defiant of their overseer. The plant has its own desires; Maund had ideas about what they were, concluding, for example, that a strong male and weak female produced better results than the reverse.[45]

For Klippart, Browne, and Maund, hybridization was a craft, not unlike that of a glassblower or a silversmith. The experimenter applied knowledge and skill to restrain the material's natural tendencies. Maund's attention to wet weather was a result of his observation that fertilization occurs in such conditions, counter to common opinion that the moisture damages the plant in blossom. He was attuned to the wheat plant's environment, properties, and sensibilities.

Farmers exercised the same awareness as they struggled to make a living in the midst of depression. Dirty, mixed, and low-quality grain hit the market, challenging the new grading systems that allowed like grain to be deposited in elevators. As grading systems became more elaborate and inspectors drew distinctions based on plumpness, purity, cleanliness, and weight, farmers charged that mixing at the boundaries yielded loss for them and profit for elevator operators and railroads. As grain became elevator receipt, new layers of knowledge and labor transformed it once more. Its cultivators resisted the transformation. Farmers learned that in regimes of measurement and classification, they invariably came up short.[46]

Currently, we read seeds through a welter of incompatible time scales: eschatological/apocalyptic, traditional/timeless, capitalistic/foreshortened. This disjointedness produces gaps in our thinking, making it difficult to understand the implications of broad social, technical, and cultural processes. We are led to believe that nature's time, in climatic or geologic terms, can give us better perspective. But historical time contains natural time, inasmuch as it renders the latter calculable through techniques of measurement and narrative.[47]

A grain of wheat tells an old story of working the earth for profit that runs counter to prevailing concepts of biological innovation and property in knowledge. Seeds are always in a state of becoming, never fixable. One can collect from the same plot of land endlessly and never get the same exact swath of material. This is simultaneously a gospel of biodiversity preservation and a point of weakness. Any gene bank, however vast, is a fragment of global biodiversity. Meanwhile, plants refuse human control more often than they accept it.

Time is not, as William Connolly has reflected, a preordained realization of an end that is already implicitly there, or a "pattern of efficient causation, determined to occur by some prior event in linear temporal order." The past remains in us, we imagine in our DNA, although the details elude us. With or without the vocabulary of functional genomics, we could also imagine time as a patch of wheat, moving not forward but inward and around, accumulating and shedding potentials.

The grammars we use to structure temporality express control, but the seed's immensity and defiance exposes their artifice, contrivance, and limits. By adjusting our perception of everyday life and cultivating moments of immersion and elongation, we create space for new ethical sensibilities and politics.[48] Maund's observation of wheat belongs in this category. So does Nassib Makarem's calligraphy, a slow and painstaking compression of all civilization onto a grain of wheat.

Makarem gave freely of his objects of art. As a member of the embattled Druze population in Syria, he first sought the favor of Ottoman governors, who ultimately disappointed his hopes for representation. He thanked the Arab nationalists who waged revolution and presented the president of France with seeds inscribed with the French national anthem. He represented Lebanon at the New York World's Fair in 1939. And he honored Henry Ford for the production of the Model T. Makarem gave his seeds to anyone he thought was forward looking, perhaps not willing to acknowledge that looking forward was a habit of history or politics rather than a virtue. As museum objects, his seeds were ultimately buried in conflict or lost in archives. But they remain an indeterminate plea for revisiting the implications of scale and scope and imagining new possibilities in a grain of wheat.

EPILOGUE : IN THE GENE BANK

When I started writing this book, I had no intention of ending it in Aleppo, Syria, either in 10,000 BCE or in the twenty-first century, but the seeds I began to observe demanded consideration as products of deep time and of global reach, subject to many temporal and geographic scales of analysis. Until recently, the International Center for Agricultural Research in the Dry Areas (ICARDA) in Tal Hadya, Syria, operated fifty kilometers southwest of Aleppo, in a region designated the heart of the Fertile Crescent, or occasionally the "Syrian Fertile Crescent," reflecting a nationalist turf war over the ownership of ancient heritage.[1]

Typically characterized as the water-rich and fertile area that was the point of origin for agriculture, the Fertile Crescent scoops across the northern Syrian desert and is bound on the east by the Tigris and Euphrates Rivers and on the west by the Mediterranean. It encompasses parts of Syria, Iraq, Lebanon, Jordan, Israel, and the West Bank in the former Mesopotamia. The majority of its marshlands are now parched in part due to drainage and damming. ICARDA's mandate is to pursue research into varieties suited for cultivation in dry areas, including those especially vulnerable to climate change; that is, not the Fertile Crescent, but rather its wreckage, the outcome of failed environmental practices and land use policies.

My first trip to ICARDA was inspired by a desire to write a better history of American wheat varieties using genetic records, and this is the project in which I am still engaged. But I returned wanting to write a better history of genetic records using American wheat varieties as well. These novel styles of knowledge and documentation powerfully structure our understandings of social and technological change, marking spaces for innovation from territory already densely populated by practices of stewardship and improvement. Technical knowledge solidifies around particular resources, which become privileged artifacts of expertise. The shedding of other forms of knowledge forces some actors to the periphery of legitimate practice while crediting others with advance. Yet nothing is lost to the extent that it cannot later be found, such that we see the rediscovery of certain agricultural practices under the rubric of sustainable development, centuries removed.

Understanding the pressures that led these practices to be abandoned serves as a corrective both to boosters of development and to critics who romanticize the agrarian past. In spite of the fashion for heritage seeds and landraces untainted by modern breeding methods, the quest for origins is in many ways misguided. As the nineteenth-century commercial origins of so many heritage seeds indicate, agricultural knowledge is characterized by mobility rather than stasis. This fluidity destabilizes property claims to seeds as material and intellectual resources, even as it challenges more appealing stories of heritage and tradition.

Rhetoric of national greatness retains a strong hold on the public imagination. Nevertheless, we are again in a global frame of mind, with new kinds of universals. Galvanized by fantasies of global connection and mobilized by awareness of anthropogenic climate change, new sciences remake seeds as new kinds of objects of research and development and potential armaments against ecological ruin.

Public and private organizations alike have rebranded seeds "plant genetic resources," and although it is a big tent, functional genomics is the favored discipline for analyzing and improving seeds. Like all disciplines, genomics is characterized by particular practices and assumptions. It is also exclusive, privileging specific forms of knowledge above others. Agronomic and farmers' knowledge have little place in disciplinary hierarchies, with the laboratory promoted over the field as a site of research.[2]

Transported from field and storage to laboratory, the seed becomes something else: a receptacle of deoxyribonucleic acid (DNA), the bases A, T, C, and G repeated over and over, coded instructions for the fabrication of life. In the vocabulary of the discipline, linguistic metaphors predominate: transcription, translation, deletion, repetition, rearrangement, and recombination. These tags and metaphors, like the Ugarit alphabet, remind us of the gaps between virtual and material life, or modes of representation and modes of being. When we regard seeds as genetic material, the code becomes indistinguishable from the matter, which is itself a grammar.

This is a poetic reduction of the seed, and not necessarily a degrading one. Put bluntly: in genetic terms wheat is smarter than you are, or at least more complicated. Common bread wheat has some 17,000,000,000 base pairs of DNA, about five times as many as people. It is a hexaploid genome, meaning it has six copies each of its seven chromosomes. Primitive wheats, along with maize, rice, and human beings, are diploid genomes, meaning they have only two copies of each chromosome. Bread wheat is complicated for other reasons as well: 80 percent of its genome consists of repetitive sequences, and the bulk is noncoding DNA. Coding DNA makes the protein sequences we know as life. The biological functions of the remainder are opaque, governing transcriptional or translational regulation of protein coding sequences and numerous imperfectly understood functions.[3]

The large size and polyploid complexity of wheat's genome have hindered genomic analysis to date. Only within the past two years have efforts been made to sequence the genome using new techniques. Even this process managed to identify only around 95,000 genes. There is nothing static about the seed we see as a result of early sequencing efforts. The genome is dynamic.[4]

Genetics is in some sense a historicist discipline, and not merely as a purveyor of new evidence of the material past.[5] Providing a set of predictable, logical, and procedural rules for inheritance, it sketches continuities across millennia while identifying fragments of what was lost. The work of bioarchaeologists and geneticists in understanding polyploid speciation casts light not simply toward a future of synthetic or transgenic wheat but also backward to the origins of our daily bread. Analysts observe that some families of genes in the Triticeae lineage associated with defense, nutritional content, energy metabolism, and growth have conserved and expanded, speculating that it is a possible result of selection during domestication. Polyploidization and domestication also resulted in the loss of many gene family members and produced many gene fragments—ancestral sequences severed over the course of centuries of breeding. But on the whole, the genome shows "extensive long-range conservation of gene order."[6] In other words, the seed's history is one of continuity and change.

But what is the historical methodology on which the sequencers rely? Analysts learn about the genome largely through comparisons with other grasses similar to wheat, diploid primitive wheats, and other cultivated grains (sorghum, rice, maize). In recent sequencing efforts, the model for comparison was patterning of single nucleotide polymorphisms (SNPs, "snips," 132,000 of them). SNPs are locations on the genome where a single "letter" of DNA varies between relatives. Maps compare sections of order between two chromosomes, in this case allowing analysis of gene conservation between wild grass and cultivated bread wheat.

These comparisons in turn provide the ground for speculations on wheat's evolution through polyploid speciation and domestication. By comparing hexaploid wheat to one of its diploid progenitors, *Aegilops tauschii*, for example, scientists can gather data on the conservation and loss of particular gene families. Losses may be magnified in modern cultivars, which make controlled single and double crosses from a uniform seed stock in lieu of simple mass selection and traditional methods of sharing seeds.

We create a past for wheat through a comparison of related contemporary genetic material. But genome sequencing can only reveal continuity or discontinuity in code: a difference or similarity between two static/contemporary sets. Any information about how change occurred is necessarily speculative. So it is not surprising that explanations for change favor models of evolution and natural selection drawn from earlier biological theories. The narrative of wheat that results is one of evolution primarily through natural selection, favoring diversification, balancing, and cyclical selection over space and time. According to one synthetic account, mutation, migration, and stochastic factors played additional roles in wheat's development.[7]

In this framing, "stochastic factors" are simply probability distributions that appear random, unintelligible according to preexisting models of polyploidization, domestication, and evolution. These sorts of evasions and dead ends increase as geneticists outsource more interpretive work to bioinformatics software devised to process large quantities of data such as that supplied by wheat's gargantuan genome. Computational solutions to problems of "big data" applied to many disciplines and objects of analysis may obscure the logic on which data is sorted and fixed into differentiated categories. Only by recovering the assumptions at the root of these tools can we understand their possibilities and limits. For if stochastic factors are simply things models cannot explain, we might also refer to them as history.

While genetics might have a historical orientation, its vocabulary and practice often serve to unwrite seeds as objects of culture, insisting instead on the arc of natural selection or the fiat of laboratory manipulation. The wheat seed, an amazing machine in its own right, is made into a neutral container for proteins assembled by technicians. Yet emerging research in epigenetics suggests that environmental and other nongenetic factors may determine the expression of genes at a cellular level, allowing for heritability of traits not altering DNA sequence—a line of reasoning some see as a return to Lamarckism, and others as an acknowledgment of natural complexity. Moreover, the human history of seeds is as much a story of power and property as it is of ecology or genetics. Mechanisms of improvement change over time, and they alter the material basis and the political economy of agricultural production in ways that shape world food systems.

Understanding the history of seed change is perhaps secondary to identifying potential applications in most genomic research projects. By identifying the genetic basis of factors such as yield, nutrient, or responsiveness to biotic and abiotic stressors, breeders can accelerate practices formerly restricted to trial and error. While genetic modification has attracted the most media attention, marker-assisted selection is as important in efforts to breed improved wheat. This practice entails the identification of microsatellite markers, or repeating sequences of base pairs of DNA, associated with wheat's key traits. In each case, key traits are open to definition, but yield generally remains primary. In the gap between imperatives of yield and the more fine-grained analysis of genetic mechanisms they justify is a dim narrative of evolution, linking polyploid speciation to human history. In turn, the genome is reread not as historical and material, but as code to be isolated and manipulated: fragments, losses, rearrangements, and recombinations.

The gulf between representation and material reproduces itself as a gulf between technical process and social/political organization. Currently, global germplasm is managed by multiple legal regimes, including the Convention on Biodiversity, the International Treaty on Plant Genetic Resources, and globally enforced intellectual property rights regulating access to seeds and plants. While multilateral agreements for sharing seeds and property rights for innovators differ in purpose, they share basic assumptions about plant genetic resources, each rooted in narratives of biological innovation supplied by mid-twentieth-century agronomists. Firstly, they provide an ordered system for naming and claiming genes, prioritizing the achievements of institutionalized research and development. Secondly, they draw sharp distinctions between traditional and modern varieties, separating seeds adapted naturally or through simple mass selection (landraces and folk varieties) from those improved by modern breeding methods (cultivars).[8] Finally, they support a proprietary conceptualization of scientific knowledge, defining property in terms of knowledge about an object rather than material possession.

These assumptions have fueled a property rights discourse that reifies scientific practice while obscuring the persistent material basis of research and the generations of knowledge on which it relies. The result is a rendering of human knowledge production that is less historical than legal and bureaucratic, providing little framework to explain incremental and collective practices of agricultural improvement or the extent to which agrobiodiversity is as much an artifact of human labor as it is a natural phenomenon.

Recent attempts to represent subaltern actors through novel species of intellectual property such as traditional knowledge (TK) and indigenous knowledge (IK) aim to value rural knowledge by converting it into property, a rearguard attempt to protect people in developing countries from uncompensated appropriations of local knowledge. Yet this parsing of knowledge as property, attributing certain ways of knowing to rural people, has the effect of denying the bearers history, creativity, and change over time. As the case of the Tatars and Mennonites demonstrates, migrants are often agents of innovation because they have been wrenched out of their geographic and cultural contexts and made to adapt to new ones. In the curiously static categories of contemporary intellectual property, people carry knowledge without making or changing it.

Current modes for biodiversity preservation and intellectual property protection cannot fully represent the people from whom seeds and knowledge are acquired. Contemporary genetic resource policies concede authority to the state rather than the embattled minorities it may harbor. Yet many populations one encounters while collecting seeds, as with the Yazidi and Kurds more broadly, are not so much traditional or indigenous as itinerant. Because of their displacement, these groups may lack political voice at national and international levels. The Kurds provide a case in point. In 1948, the USDA agronomist Jack Harlan collected wheat from Kurdish farmers who had migrated from northern Iraq to eastern Turkey, taking their seeds with them.[9] The improved variety that made use of this landrace became the most widespread variety in the United States in the 1950s and 1960s. But, as the Yazidi herder would be the first to observe, Iraq, Syria, and Turkey did not officially recognize their Kurdish populations. And today each country still retains its own genetic resource policy.

How do collecting expeditions and the research associated with them rely on, challenge, or circumvent geopolitical networks and relationships? Who benefits from multilateral treaties for sharing plant genetic resources, and in what respects? Who resists them, and why? What would a plant genetic resource policy look like that was oriented toward producers rather than nation-states? Or that abandoned its romance with tradition in favor of a more pragmatic and realist orientation toward change over time?

These are questions that require us to look beyond old frameworks of national histories and area studies toward new geographies that recognize networks of research capital as primary. Existing orders may be products of empire, yet they are not identical to their antecedents, and understanding their operations requires approaches that eschew the nation-state and its archives as primary movers. Old categories of center and periphery obscure new networks of exchange intertwined with imperial legacies of removal and settlement, including the politics of resource exploitation and border creation. To collectors, at times the entire landscape seems stitched in jagged seams.

Waterways are sites of conflict, with hydroelectric projects threatening to reduce the Aras River (or the Tigris and Euphrates) to a trickle. Meanwhile, plants separated by several meters and a national border may be regarded as global commons on one side and national patrimony on the other, with collection of the latter restricted by national and international accords protecting endangered species.

Seeds are powerful signifiers because they compress future potential and deep past into objects both minuscule and abundant. "I have great faith in a seed. Convince me that you have a seed there, and I am prepared to expect wonders," wrote Henry David Thoreau, making the wonder stand in for a multitude of biological processes. A seed is "an orchard invisible," goes the Welsh proverb, conjuring up futurity in the present tense using spatial imagery of multiplication.[10] The seed is a metaphor for the potential of all things to come, imagined as constant ongoing and ceaseless growth.

Yet seeds are also an embodiment of things past. The agronomist Cecil Salmon, who conveyed Norin 10 from Japan to the United States in 1945, described nineteenth-century seed varieties as an "ancestral swamp" upon which modern plant breeding efforts have been based. The word "swamp," however, is a pejorative term to describe what is in fact a complex ecosystem. And the same could be said of Salmon's characterization of plant genetic resources in nineteenth-century America, which were the products of generations of stewardship and creative labor. Institutionalized biological innovation is neither linear nor progressive, but rather a process of collection, appropriation, and organization ritually repeated by amassing genetic resources and naming or classifying them according to new schema that render their human provenance secondary.

In the void of meaningful action at the human scale, we see the return of sacred temporalities to justify breeding efforts and world biodiversity preservation. Resurrecting Malthusian arguments, biotech companies tell us we must increase yields to feed the world's growing population or face collapse. In response the crisis of the Anthropocene, preservationists advocate seed vaults as safeguards against extinction. Prophecy works, as Reinhart Koselleck observed, by destroying time through its fixation on the End. Unlike prognosis, "it renews itself on the absence of what is predicted."[11] The recourse to apocalypse allows prophets to justify action under the cover of preventing extinction. We should regard appeals to the global with skepticism when they insinuate commonality and universality to circumvent political action rather than engaging it.

In the twenty-first century, political futures remain pegged to the success of agriculture and rural development. Regions afflicted by conflict and environmental crisis are especially vulnerable to the reconstructions of national governments and international aid agencies mired in ideological projections of agriculture and development. Too often the vocabularies of biodiversity, biotechnology, and development employed at national and international levels represent rural people and agrarian knowledge as changeless and primitive, awaiting modernization. These characterizations of people as passive objects of improvement draw attention from the more immediate legacies of conflict and crisis crippling local economies.

Efforts to reform these practices will fail as long as they rely on formulaic histories of economic change that misrepresent how knowledge is made, especially toward the creation of markets of scale. Agricultural expansion in the United States, rather than an effervescence of innovation, was a muddled and circuitous practice of accumulation, rebranding, and reorganization of diverse intellectual and material resources in new institutions of commerce and governance. These consolidations masked the improvisational, incremental, ephemeral, and migratory qualities of knowledge across time. In their deep timescales, incessant regeneration, and defiance of human control, seeds challenge us to reconsider linear or progressive models of innovation, and of history as a form of storytelling about the past.

What kinds of stories should we tell about the past in a moment of renewed awareness that we inhabit a global system? Arguably, concepts of individuality and agency derived from political theory and natural science have reached their limits as ways of explaining the world, along with an economic system that has for five centuries, as Jason Moore has recently argued, relied on "Cheap Nature" strategies that extract and exploit labor power, energy, food, and raw materials.[12] Even as chronic economic and environmental crises have revealed the corruption of these strategies, environmental exploitation for profit has escalated. Inequality is persistent, species extinction possible. In the darkest estimation, the apocalypse has already happened: its effects are just unequally distributed.[13] Following Anna Tsing, maybe we should realize ourselves as casualties of the economic processes in which we participate and look to excreta of global commodity chains for insight into life among the ruins.[14] Or as Natasha Myers has argued: plants witness us, and we can learn from their difference.[15] And perhaps we can also learn from the varied ways people of the past have regarded them.

Innovation is one of many myths for understanding the interdependence of human and nonhuman actors. There are others. In an Osage creation story, the earth was first covered with water, and the people, born of the sun and the moon, floated helplessly in the air with the animals. They asked the elk for help because of his nobility and trustworthiness. As he lowered himself into the water and began to sink, he called the winds to blow in all directions. As wind carried the water upward, at first only rocks appeared, where nothing would grow. But when finally soft earth became visible, the elk, overcome with joy, rolled and rolled on his back. His loose hairs clung to the ground he disturbed and grew there, producing beans, maize, potatoes, wild turnips, and finally grasses and trees. People wandered over the uncovered earth. When they found footsteps, they followed them and joined one another, traveling in search of food.[16] Their footprints traced a history of migration in which human settlers were students rather than masters of their environments, grateful to a contented elk for rolling in the mud.

ACKNOWLEDGMENTS

The writing of this book has been a study in detours, and I have appreciated the people I met at every turn. Grants from the Whiting Foundation, Andrew W. Mellon Foundation, National Science Foundation, German Academic Exchange Service (DAAD), and the Colonel Return Jonathan Meigs the First Fund at Wesleyan University provided resources for me to pursue this project. Fellowships from the New-York Historical Society, Smithsonian Institution, Hagley Museum and Library, Library Company of Philadelphia, and American Philosophical Society enabled research in their collections. I owe special thanks to the archivists and librarians of these institutions, as well as those of the Kansas State Historical Society, Lloyd Museum and Library, and Wesleyan University.

A fellowship at the Rutgers Center for Historical Analysis in the program on Networks of Exchange was indispensable to the completion of this book, which is better for the engagement of program directors Toby Jones and James Delbourgo and fellows Chris Blakley, Eli Cook, Julia Fein, AJ Murphy, Juno Salazar Parreñas, and Jamie Pietruska. One also learns through absence, and while in my office at 88 College Avenue I turned frequently to the writings of Philip Pauly and Susan Schrepfer.

I owe special thanks to Dan Bouk, Ann Fabian, and Emily Pawley, as well as my reviewers at the University of Chicago Press, for helpful comments and interventions on the complete manuscript. Catherine Evtuhov, Michael Flannery, Richard John, Martin Jones, and John Staples generously offered detailed comments on individual chapters.

Workshops and colloquia provide the ground for ideas in the making, and the source of subsequent collaborations. At Columbia, the biweekly colloquium of the Mellon Institute for Social and Economic Research and Policy (ISERP) Graduate Program, directed by William McAllister, provided a space to think across the humanities and social sciences. The Yale workshop Plants, Animals, and Ownership with Daniel Kevles, Jean-Paul Gaudillière, Helen Curry, Christophe Bonneuil, and others provided important insight at an early stage of my research, as did readings by Roger Horowitz, Susan Strasser, and Phil Scranton at the Hagley Center for the History of Business and Technology. Colleagues at the Max Planck Institute for the History of Science encountered my new research on plant genetic resource management at its most inchoate, and I thank Elena Aronova, Etienne Benson, Lorraine Daston, David Sepkosi, and Christine von Oertzen for perspicacious comments and questions. More recently, I've benefited from participation in a Yale University symposium, "Grassroots Modernities: Nature, Culture, and Improvement in the Atlantic World," organized by Ariel Ron and Emily Pawley, whose reflections on our shared interests have helped sustain my research.

As a teacher, I emulate the persistent encouragement and critique I received from Eric Foner and Elizabeth Blackmar at Columbia University, along with the intellectual generosity of Casey Blake, Barbara Fields, Matthew Jones, Alice Kessler-Harris, Pamela Smith, and Anders Stephanson. At the Smithsonian and the Hagley, respectively, I benefited from Pamela Henson's encyclopedic knowledge of the Smithsonian and Roger Horowitz's expansive readings in the history of business and technology. I learned from our conversations and their careful readings.

My colleagues in the departments of History, Environmental Studies, and Science in Society at Wesleyan University have inspired me to think in new ways about the past. I owe special thanks to Demetrius Eudell, Patricia Hill, William Pinch, Ron Schatz, Gary Shaw, Magda Teter, and Jennifer Tucker for readings of my work. I have benefited from Barry Chernoff, Joseph Rouse, and Jill Morawski's commitment to the interdisciplinary study of science and environment. Jill Morawski's direction of the Center for the Humanities at Wesleyan made my semester fellowship in residence there a generative space for interdisciplinary research.

Perhaps the most unexpected part of this journey has been learning to see the world through its flora. In my collaborations with plant genetic resource specialists, I have learned an enormous amount from Ahmed Amri, Josephine Piggin, Sergey Shuvalov, Natalya Rukhyan, and Zane Webber. They have been my teachers and my traveling companions in some extraordinary environments.

At the University of Chicago Press, Karen Darling and Evan White have shepherded this book into being with patience and enthusiasm, without which there would be no book. Finally, ideas are made in conversation with others, and I'm grateful for conversations in Berlin with Mirjam Brusius, Anne Dippel, Yulia Frumer, Megan McNamee, Isabell Schrickel, Alma Steingart, Kathleen Vongasthorn, Florian Wallner, Christina Vagt, and Oriana Walker, and stateside with Elizabeth Bouk, Sonali Chakravarti, Deborah Coen, Jefferson Decker, Konstantin Dierks, Paul Erickson, Matthew Fetchko, David Huyssen, Amy Offner, Jason Petrulis, Seth Rockman, David Singerman, Bonnie Smith, Victoria Smolkin, Laura Stark, Laura Ann Twagira, Amrys Williams, Rebecca Woods, and Anya Zilberstein—and so many others that to name them would be an encyclopedic project on par with the *Species plantarum*. I owe thanks to these people for alternately reading, asking about, and not asking about my work. On that score, no one deserves more thanks than my family, and especially my parents Tom and Donna Fullilove, who cared for me even when I resembled a plant that had not been watered for one too many days. I learned from this example most of all.

NOTES

PROLOGUE

1. A landrace is defined as a regional ecotype, locally adapted variety, or traditional variety of a domesticated species of plant or animal, generally distinguished by its isolation from other populations of the species. It is typically opposed to a cultivar, which is a variety produced by selective breeding and maintained by propagation. The landrace concept has been called into question in part because of the hard line it draws between laboratory-based breeding and farmer selection. See, for example, Trygve Berg, "Landraces and Folk Varieties: A Conceptual Reappraisal of Terminology," *Euphytica* 166, no. 3 (2009): 423–30.

2. Warren C. Baum, Michael L. Lejeune, and World Bank, *Partners against Hunger: The Consultative Group on International Agricultural Research* (Washington, DC: Published for the CGIAR by the World Bank, 1986).

3. For example, Jared M. Diamond, *Guns, Germs, and Steel: The Fates of Human Societies* (New York: W. W. Norton, 1998); Alfred W. Crosby, *The Columbian Exchange: Biological and Cultural Consequences of 1492* (Westport, CT: Greenwood Press, 1972); William Hardy McNeil, *Plagues and Peoples* (New York: Doubleday, 1989).

4. Several recent histories of energy regimes are Andreas Malm, *Fossil Capital: The Rise of Steam-Power and the Roots of Global Warming* (London: Verso, 2016); Christopher F. Jones, *Routes of Power: Energy and Modern America* (Cambridge, MA: Harvard University Press, 2014); Sean P. Adams, *Old Dominion, Industrial Commonwealth: Coal, Politics, and Economy in Antebellum America* (Baltimore: Johns Hopkins University Press, 2004); Richard White, *The Organic Machine* (New York: Hill and Wang, 1995).

5. John Bellamy Foster has described this extractive agriculture as a "metabolic rift" between humans and their environments, suggesting that mid-nineteenth-century preoccupation with overpopulation and soil infertility led Karl Marx and others to be concerned about the unsustainability of capitalist agriculture, and human estrangement from the natural world. John Bellamy Foster, *Marx's Ecology Materialism and Nature* (New York: Monthly Review Press, 2000).

6. On plant movements and human migration, Crosby, *Columbian Exchange*; Alfred W. Crosby, *Ecological Imperialism: The Biological Expansion of Europe, 900–1900* (New York: Cambridge University Press, 1986). Judith Carney has revised Crosby's rendering of the Americas as neo-Europes by focusing on the purposeful and incidental transplantation of African crops and agricultural knowledge. Judith Carney and Richard Rosomoff, *In the Shadow of Slavery: Africa's Botanic Legacy in the New World* (Berkeley: University of California Press, 2011).

7. On European botanic gardens and tropical agriculture, Richard Harry Drayton, *Nature's Government: Science, Imperial Britain, and the "Improvement" of the World* (New Haven, CT: Yale University Press, 2000); E. C. Spary, *Utopia's Garden: French Natural History from Old Regime to Revolution* (Chicago: University of Chicago Press, 2000); Richard Grove, *Green Imperialism: Colonial Expansion, Tropical Island Edens, and the Origins of Environmentalism, 1600–1860* (New York: Cambridge University Press, 1995). On paper technologies as the infrastructure of collecting, Jacob Soll, "From Note-Taking to Data Banks: Personal and Institutional Information Management in Early Modern Europe," *Intellectual History Review* 20, no. 3 (2010): 355–75; Staffan Müller-Wille and Sara Scharf, *Indexing Nature: Carl Linnaeus (1707–*

1778) and His Fact-Gathering Strategies, 2009, http://www2.1se.ac.uk/economicHistory/pdf/FACTSPDF/3909MuellerWilleScharf.pdf. On practices and institutions of botanical collection, for example: Harold John Cook, *Matters of Exchange: Commerce, Medicine, and Science in the Dutch Golden Age* (New Haven, CT: Yale University Press, 2007); Paula Findlen, *Possessing Nature: Museums, Collecting, and Scientific Culture in Early Modern Italy* (Berkeley: University of California Press, 1994); Londa L. Schiebinger, *Plants and Empire: Colonial Bioprospecting in the Atlantic World* (Cambridge, MA: Harvard University Press, 2004); Jim Endersby, *Imperial Nature: Joseph Hooker and the Practices of Victorian Science* (Chicago: University of Chicago Press, 2008).

8. On four stages theory, Drew R. McCoy and Institute of Early American History and Culture (Williamsburg, Virginia), *The Elusive Republic: Political Economy in Jeffersonian America* (Chapel Hill: Published for the Institute of Early American History and Culture, Williamsburg, Virginia, by the University of North Carolina Press, 1980). On Jefferson's fears of colonial degeneracy, Philip J. Pauly, *Fruits and Plains: The Horticultural Transformation of America* (Cambridge, MA: Harvard University Press, 2007), 10–32.

9. William Cronon, *Nature's Metropolis: Chicago and the Great West* (New York: W. W. Norton, 1991); Margaret W. Rossiter, *The Emergence of Agricultural Science: Justus Liebig and the Americans, 1840–1880* (New Haven, CT: Yale University Press, 1975); Deborah Kay Fitzgerald, *Every Farm a Factory: The Industrial Ideal in American Agriculture* (New Haven, CT: Yale University Press, 2003); Jack Ralph Kloppenburg, *First the Seed: The Political Economy of Plant Biotechnology, 1492–2000* (Madison: University of Wisconsin Press, 2004); Alan L. Olmstead and Paul Webb Rhode, *Creating Abundance: Biological Innovation and American Agricultural Development* (New York: Cambridge University Press, 2008); Pauly, *Fruits and Plains*; Donald Worster, *Under Western Skies: Nature and History in the American West* (Oxford: Oxford University Press, 1992).

10. In the past decade, historians have turned their attention to the lived experience of agricultural settlement, and how settlers understood land, environment, and production during periods of rapid change: Joyce E. Chaplin, *An Anxious Pursuit: Agricultural Innovation and Modernity in the Lower South, 1730–1815* (Chapel Hill: University of North Carolina Press, 1993); Benjamin R. Cohen, *Notes from the Ground: Science, Soil, and Society in the American Countryside* (New Haven, CT: Yale University Press, 2009); Emily Pawley, "'The Balance-Sheet of Nature': Calculating the New York Farm, 1820–1860" (PhD diss., University of Pennsylvania, 2009), http://repository.upenn.edu/dissertations/AAI3381774; Steven Stoll, *Larding the Lean Earth: Soil and Society in Nineteenth-Century America* (New York: Hill and Wang, 2002); Conevery Bolton Valenčius, *The Health of the Country: How American Settlers Understood Themselves and Their Land* (New York: Basic Books, 2002).

11. For example, Nick Cullather, *The Hungry World: America's Cold War Battle against Poverty in Asia* (Cambridge, MA: Harvard University Press, 2010); James McCann, *Maize and Grace: Africa's Encounter with a New World Crop, 1500–2000* (Cambridge, MA: Harvard University Press, 2005). Extending foundational work in development studies, e.g., Arturo Escobar, *Encountering Development: The Making and Unmaking of the Third World* (Princeton, NJ: Princeton University Press, 1995).

12. As innovation, Olmstead and Rhode, *Creating Abundance*; as exploitation, Sven Beckert, *Empire of Cotton: A Global History* (New York: Alfred A. Knopf, 2014); Edward E. Baptist, *The Half Has Never Been Told: Slavery and the Making of American Capitalism* (New York: Basic Books, 2014). Ian Beamish articulated this polarity effectively at the Yale Center for Representative Institutions colloquium "Grassroots Modernities: Nature, Culture, and Improvement in

the Atlantic World," June 9, 2015. Beckert's is one recent attempt to shift histories of capitalism toward rural sites of production.

13. The history of twentieth-century agro-biodiversity preservation initiatives is still to be written. One recent survey is Christophe Bonneuil and Marianna Fenzi, "From 'Genetic Resources' to 'Ecosystems Services': A Century of Concerns and Global Policies to Conserve Crop Diversity," forthcoming in *Culture, Agriculture, Food, and Environment.* See also Robin Pistorius, *Scientists, Plants, and Politics: A History of the Plant Genetic Resources Movement* (Rome: International Plant Genetic Resources Institute, 1997); Paul Gepts et al., *Biodiversity in Agriculture: Domestication, Evolution, and Sustainability* (New York: Cambridge University Press, 2012), section 4; Stephen B. Brush, *Farmers' Bounty: Locating Crop Diversity in the Contemporary World* (New Haven, CT: Yale University Press, 2004), chap. 8.

14. Two perspectives: Miguel A. Altieri, *Agroecology: The Science of Sustainable Agriculture* (Boulder, CO: Westview Press, 1995); Raj Patel, "The Long Green Revolution," *Journal of Peasant Studies* 40, no. 1 (2013): 1–63.

15. Karl Jacoby, *Crimes against Nature: Squatters, Poachers, Thieves, and the Hidden History of American Conservation* (Berkeley: University of California Press, 2001); Richard West Sellars, *Preserving Nature in the National Parks: A History* (New Haven, CT: Yale University Press, 1997).

16. On the deployment of the vocabulary and food security in international relations during World War II and its subsequent career in the UN, Mark Mazower, *Governing the World: The History of an Idea, 1815 to the Present* (New York: Penguin, 2013), 191–213; Raj Patel, guest editor, "Food Sovereignty," *Journal of Peasant Studies* 36, no. 3 (2009): 663–706. On natural economy, e.g., E. C. Spary, "Political, Natural, and Bodily Economies," in *Cultures of Natural History*, ed. N. Jardine, J. A. Secord, and E. C. Spary (Cambridge: Cambridge University Press, 1996), 178–96; Pauly, *Fruits and Plains*, 10–32.

17. Robert H. Wiebe, *The Search for Order, 1877–1920* (New York: Hill and Wang, 1967).

18. Ecclesiastes 5:8–9, ESV.

FIELD NOTES: "GREEN REVOLUTIONS"

1. Two English-language biographies of Vavilov are Igor G. Loskutov and International Plant Genetic Resources Institute, *Vavilov and His Institute: A History of the World Collection of Plant Genetic Resources in Russia* (Rome: International Plant Genetic Resources Institute, 1999), and Peter Pringle, *The Murder of Nikolai Vavilov: The Story of Stalin's Persecution of One of the Great Scientists of the Twentieth Century* (New York: Simon and Schuster, 2008).

2. On Vavilov as modernizer, Bonneuil and Fenzi, "From 'Genetic Resources' to 'Ecosystems Services.'" On Russian biology and the eugenics movement, Mark B. Adams, *The Wellborn Science: Eugenics in Germany, France, Brazil, and Russia* (New York: Oxford University Press, 1990), 153–70.

3. L. P. Reitz and S. C. Salmon, "Origin, History, and Use of Norin 10 Wheat," *Crop Science* 8, no. 6 (1968): 686–89.

4. Cullather, *Hungry World*; John H. Perkins, *Geopolitics and the Green Revolution: Wheat, Genes, and the Cold War* (New York: Oxford University Press, 1997). The Italian wheat breeder Nazareno Strampelli worked with Japanese dwarf varieties in the 1910s, producing varieties cultivated in Italy and Argentina in the interwar period. After World War II, breeders in Eastern bloc countries (Yugoslavia, Hungary, Bulgaria, Romania, and Czechoslovakia) imported Italian varieties. In the USSR, breeders at the Krasnodar experiment station developed Bezostaya, which also had dwarfing genes derived from Strampelli material. Bezostaya became the most common Soviet variety cultivated in the second half of the twentieth century. On the parallel

pathways of semidwarf wheat research, see K. Borojevic and K. Borojevic, "The Transfer and History of 'Reduced Height Genes' (Rht) in Wheat from Japan to Europe," *Journal of Heredity* 96, no. 4 (2005): 455–59.

5. On the ancestry of Norin 10 and Turkey wheat, see K. S. Quisenberry and L. P. Reitz, "Turkey Wheat: The Cornerstone of an Empire," *Agricultural History* 48, no. 1 (1974): 98–110.

6. David Moon, "In the Russians' Steppes: The Introduction of Russian Wheat on the Great Plains of the United States of America," *Journal of Global History* 3, no. 2 (2008): 203–25.

7. Bonneuil and Fenzi, "From 'Genetic Resources' to 'Ecosystems Services.'"

8. Food and Agriculture Organization of the United Nations, *FAO Statistical Yearbook: World Food and Agriculture* (Rome: FAO, 2013), http://www.fao.org/docrep/018/i3107e/i3107e00.htm.

CHAPTER ONE

1. James Morrow and Allan B. Cole, eds., *A Scientist with Perry in Japan: The Journal of Dr. James Morrow* (Chapel Hill: University of North Carolina Press, 1947), 6.

2. Ibid., 42–44.

3. Ibid., 260. Asa Gray's substantial delays in identifying the specimens earned Perry's ire and delayed their inclusion in the official narrative of the expedition. See *Narrative of the Expedition of the American Squadron to Japan*, 1856, correspondence on pp. 299 ff., followed by Gray's *Account of the Botanical Specimens*.

4. Alfred Hunter, *A Popular Catalogue of the Extraordinary Curiosities in the National Institute, Arranged in the Building Belonging to the Patent Office* (Washington, DC: A. Hunter, 1857), 58–9. There are two plowshares from the Perry Expedition in the Smithsonian's collections and the ledger of the US National Museum, which took custody of the government science collections housed in the upper galleries of the Patent Office in 1857: Chang-su Houchins, *Artifacts of Diplomacy: Smithsonian Collections from Commodore Matthew Perry's Japan Expedition (1853–1854)* (Washington, DC: Smithsonian Institution Press, 1995), 123.

5. For example, Lewis Henry Morgan's thirty-five years of study on Indian material culture and kinship culminated in the publication of *Ancient Society* (New York, 1877). Morgan linked phases of civilization to technological development and social institutions such as property, drawing on Henry James Sumner Maine's theories of law as developing "from status to contract." Henry James Sumner Maine, *Ancient Law: Its Connection with the Early History of Society, and Its Relation to Modern Ideas* (London, 1861). On theories of social evolution, see George Stocking, *Victorian Anthropology* (New York: Free Press, 1991); Curtis Hinsley, *Savages and Scientists: The Smithsonian Institution and the Development of American Anthropology, 1846–1910* (Washington, DC: Smithsonian Books, 1981); Robert Bieder, *Science Encounters the Indian, 1820–1880: The Early Years of American Ethnology* (Norman: University of Oklahoma Press, 1989); Alan Joyce, *The Shaping of American Ethnography: The Wilkes Exploring Expedition, 1838–1842* (Lincoln: University of Nebraska Press, 2001).

6. On the social status of naturalists, see Anne Secord, "Artisan Botany," in *Cultures of Natural History*, ed. N. Jardine, J. A. Secord, and E. C. Spary (Cambridge: Cambridge University Press, 1996), 378–93; on local practices of classification and naming, e.g., Alix Cooper, *Inventing the Indigenous: Local Knowledge and Natural History in Early Modern Europe* (New York: Cambridge University Press, 2007).

7. This is Philip Pauly's estimation in *Fruits and Plains*, 104–5.

8. Douglas Evelyn, "The National Gallery at the Patent Office," in *Magnificent Voyagers: The U.S. Exploring Expedition, 1838–1842*, ed. Herman Viola and Carolyn Margolis (Washington, DC: Smithsonian Institution Press, 1985).

9. On collections as representations of expansionism, Curtis Hinsley, *Savages and Scientists: The Smithsonian Institution and the Development of American Anthropology, 1846–1910* (Washington, DC: Smithsonian Institution Press, 1981); William H. Goetzmann, *Exploration and Empire: The Explorer and the Scientist in the Winning of the American West* (Austin: Texas State Historical Association, 1994).

10. "Position, Character, and Aspiration of the National Institute," 1843; John M. Porter, 1842, and November 30, 1843, letters to the National Institute, Richard Rathbun Papers, National Institute, Smithsonian Institution Archives.

11. Vincent Ponko Jr., *Ships, Seas, and Scientists: US Naval Exploration and Discovery in the Nineteenth Century* (Annapolis, MD: Naval Institute Press, 1974).

12. Ivan Karp et al., eds., *Museum Frictions: Public Cultures/Global Transformations* (Durham, NC: Duke University Press, 2006), 1–11.

13. Postcolonial theorists emphasize the extent to which national museums are products of imperialism. For example, Tony Bennett, *The Birth of the Museum: History, Theory, Politics* (New York: Routledge, 1995); Homi K. Bhabha, *The Location of Culture* (New York: Routledge, 1994); and Homi K. Bhabha's review essay of the exhibition *Circa 1492: Art in the Age of Exploration* at the National Gallery in Washington, DC: "Double Visions," in *Grasping the World: The Idea of the Museum*, ed. Donald Preziosi and Claire Farago (Burlington, VT: Ashgate, 2004). On the relevance of these circuits of knowledge, material, and people to North American history, Ann Laura Stoler, ed., *Haunted by Empire: Geographies of Intimacy in North American History*, American Encounters/Global Interactions (Durham, NC: Duke University Press, 2006).

14. While museums may be places where disputes over knowledge and institutional claims resolve, failure to resolve is as historically significant. The museum in the Patent Office is illegible according to most models of state power, from Weber's bureaucracy to James Scott's omnipotent state to Foucault's governmentality: James Scott, *Seeing Like a State: How Certain Schemes to Improve the Human Condition Have Failed* (New Haven, CT: Yale University Press, 1998); Michel Foucault, "Governmentality," trans. Rosi Braidotti and revised by Colin Gordon, in *The Foucault Effect: Studies in Governmentality*, ed. Graham Burchell, Colin Gordon, and Peter Miller (Chicago: University of Chicago Press, 1991), 87–104; Sophie Forgan, "Building the Museum: Knowledge, Conflict, and the Power of Place," *Isis* 96 (2005): 572–85. On museums as legitimating national or imperial governance, Benedict Anderson, *Imagined Communities* (London: Verso, 1983); Bernard S. Cohn, *Colonialism and Its Forms of Knowledge: The British in India* (Princeton, NJ: Princeton University Press, 1996). William Novak and others have debunked the myth of the weak state in the nineteenth-century United States, but the haphazardness of its development warrants further analysis. On the myth of the weak state, W. J. Novak, "The Myth of the 'Weak' American State," *American Historical Review* 113, no. 3 (2008): 752–72.

15. Richard Rathbun and Smithsonian Institution, *The National Gallery of Art: Department of Fine Arts of the National Museum* (Washington, DC: Government Printing Office, 1916), 25–44.

16. National Museum of American History (US), Division of Political History, and William L Bird, *Souvenir Nation: Relics, Keepsakes, and Curios from the Smithsonian's National Museum of American History* (New York: Princeton Architectural Press, 2013), 19–27.

17. Manufacturers studied models to improve their machinery, the purpose for which the hall was ostensibly established. C. M. Harris, "Specimens of Genius and Knickknacks: The Early Patent Office and Its Museum," *Prologue: The Journal of the National Archives* 23 (1991): 406–17. Robert P. Multauf, "Catalogue of Instruments and Models in the Possession of the

American Philosophical Society" (Philadelphia, 1961); Eugene S. Ferguson, "Technical Museums and International Exhibitions," *Technology and Culture* 6 no. 1 (1965): 30–46; Charles R. Richards, *The Industrial Museum* (New York: Macmillan, 1925); David J. Jeremy, *Transatlantic Industrial Revolution: The Diffusion of Textile Technologies between Britain and America, 1790–1830s* (Oxford: Basil Blackwell, 1981).

18. "National Institute, 1850, 1851, 1852, 1853, 1854," Richard Rathbun Papers, The National Institute for the Promotion of Science—Draft MS, Notes, and Research Material, RG 7078, NI, SIA, Washington, DC.

19. Perry and his officers erroneously believed they negotiated with the emperor rather than the shōgun. Perry's mistaken contention that the emperor was the audience for his letter conceals the extent to which the liberalization of trade relations was a by-product of the disintegration of the Tokugawa shogunate rather than his personal achievement. For a summary of the Perry expedition's objectives, Morrow and Cole, *Scientist with Perry*, x–xiii. The Americans were in competition with Russians and British envoys to establish trade relations with the Japanese: see, e.g., William McOmie, *The Opening of Japan, 1853–1855: A Comparative Study of the American, British, Dutch, and Russian Naval Expeditions to Compel the Tokugawa Shogunate to Conclude Treaties and Open Ports to Their Ships* (Folkestone, England: Global Oriental, 2006). The principal source of information about the expedition is the official narrative, United States Congress, *Narrative of the Expedition of an American Squadron to the China Seas and Japan, Performed in the Years 1852, 1853, and 1854, under the Command of Commodore M. C. Perry, United States Navy, by Order of the Government of the United States. Compiled from the Original Notes and Journals of Commander Perry and His Officers, at His Request, and under His Supervision, by Francis L. Hawks, D.D.L.L.D. with Numerous Illustrations* (Washington, DC, 1855) and vol. 2 of same, 1856.

20. Morrow and Cole, *Scientist with Perry*, 5, 10, on Buist; 155, on Landreth; D. Landreth Seed Company Newsletter (D. Landreth Seed Co., New Freedom, Pa.), June 2009, vol. 205, issue 206, available at http://www.landrethseeds.com/newsletters/; D. Landreth Seed Co. Historical Timeline, http://www.saveseeds.org/biography/landreth/landreth_timeline.html. I am grateful to Daniel Kevles for the information on Landreth.

21. Morrow and Cole, *Scientist with Perry*, xxvii, 255–57.

22. Ibid., 11–23.

23. Perry to Secretary of Navy, Madeira, December 14, 1852, regarding ports refuge, *Narrative of the Expedition*, 85–87.

24. Bayard Taylor and surgeon C. F. Fahs issued "Report on the Exploration of Peel Island" as part of the official narrative, as did Captain Joel Abbot for the Bonin Islands as a group: *Narrative, 1855*, 67–78, 125–33. After Taylor and Fahs's initial reports, Perry turned his attention to the Bonin Islands as a possible port of refuge: *Narrative, 1855*, 209–12. It's difficult to know why Perry initially identified Lew Chew/Okinawa as a possible base rather than the Bonin Islands, which had already been surveyed by British expeditions, e.g., G. Tradescant Lay, *Trade with China: A Letter Addressed to the British Public on Some of the Advantages That Would Result from an Occupation of the Bonin Islands* (London: Royston and Brown, 1837). On a subsequent trip Perry had garden seeds of every description "distributed among the present settlers, and hopes were held out to them by the Commodore of a future supply of implements of husbandry and a greater number of animals." *Narrative of the Expedition, Volume II, 1856*, 150 ff.

25. Morrow and Cole, *Scientist with Perry*, 107–9.

26. Ibid., 91.

27. Ibid., 96–100.

28. Ibid., 111. James Morrow et al., Correspondence, 1860–65, James Morrow Papers, 1838–1938, University of South Carolina Library, Columbia.

29. Morrow and Cole, *Scientist with Perry*, 42–44.

30. Ibid., 175–76, 199.

31. Ibid., 151–52.

32. Ibid.

33. William Speiden, *With Commodore Perry to Japan: The Journal of William Speiden Jr., 1852–1855* (Annapolis, MD: Naval Institute Press, 2013), July 16; *Narrative*, chap. 14.

34. Morrow and Cole, *Scientist with Perry*, 151–52.

35. S. Wells (Samuel Wells) Williams, *A Journal of the Perry Expedition to Japan (1853-1854)* (Wilmington, DE: Scholarly Resources, 1973), 159.

36. Morrow and Cole, *Scientist with Perry*, 152.

37. Ibid., 125; Wilhelm Heine, *With Perry to Japan: A Memoir* (Honolulu: University of Hawai'i Press, 1990), 208.

38. Morrow and Cole, *Scientist with Perry*, 170–75.

39. Ibid., 187.

40. Ibid., 170–76.

41. John Varden, "John Varden's Museum," [1856?], "MS A," John Varden Papers, RU 7063, Box 1, SIA.

42. National Institute Memorial to Congress, November 14, 1848, Richard Rathbun Papers, NI, SIA.

43. Ibid.

44. Anne Secord characterizes nature collection in the anthropological vocabulary of gift exchange: Anne Secord, "Corresponding Interests: Artisans and Gentlemen in Nineteenth-Century Natural History," *British Journal for the History of Science* 27 no. 4 (1994): 383–408. On collectors' networks, see also Daniel Goldstein, "'Yours for Science': The Smithsonian Institution's Correspondents and the Shape of Scientific Community in Nineteenth-Century America," *Isis* 85 (December 1994): 573–99; Susan Scott Parish, *American Curiosity: Cultures of Natural History in the Colonial British Atlantic World* (Chapel Hill: University of North Carolina Press, 2006); on Bartram, Charlotte M. Porter, "Natural History Discourse and Collections: The Roles of Collectors in the Southeastern Colonies of North America," *Museum History* 1 (2007). On commerce and rights in transacted specimens, see Janet Browne, "Do Collections Make the Collector? Charles Darwin in Context," in *From Private to Public*, ed. Marco Beretta (Uppsala: Watson, 2005).

45. "Position, Character, and Aspiration of the National Institute," 1843, Richard Rathbun Papers, NI, SIA.

46. According to its final memorial to Congress, "many institutions, learned bodies, societies, and individuals at home and abroad" were eager to procure the duplicate specimens, a wish that should be granted for the sake of science and mutual enrichment. National Institute Memorial to Congress, November 14, 1848, Richard Rathbun Papers, NI, SIA.

47. Ibid.

48. S.Doc 95, 25th Congress, 3rd Session, January 28, 1839. "Letter from the Commissioner of Patents to the Chairman of the Committee on Patents and the Patent Office, in relation to the collection and distribution of seeds and plants."

49. Pauly, *Fruits and Plains*, 104–5.

50. See A. Hunter Dupree, *Science and the Federal Government* (Baltimore: Johns Hopkins University Press, 1986), and Sally Kohlstedt, "A Step toward Scientific Self-Identity in the

United States: The Failure of the National Institute, 1844," *Isis* 62 (1971): 339, the latter of which discusses the competition of the Association of American Geologists and Naturalists for scientific authority.

51. The Records of the National Institute, SIA, contain numerous letters regarding Tappan's liberation of specimens for himself and his connections. Ultimately, museum custodian John Varden made written notations whenever books or materials were borrowed from the offices and halls of the US Ex. Ex. collections, but Varden was a poor record keeper and these survive as scattered bits of paper rather than as a single ledger.

52. On the life and death of Ro-Veidovi, Ann Fabian, *The Skull Collectors: Race, Science, and America's Unburied Dead* (Chicago: University of Chicago Press, 2010).

53. Kristian Kristiansen, "The Destruction of Archaeological Heritage and the Formation of Museum Collections," in *Learning from Things: Method and Theory of Material Culture Studies*, ed. David W. Kingery (Washington, DC: Smithsonian Institution Press, 1996).

CHAPTER TWO

1. Agricultural societies and journals agitated for the elevation of the Patent Office Agricultural Department to agency level, while common farmers remained resentful of elite pretensions, educational prescriptions, and government taxes. Margaret Rossiter, *The Emergence of Agricultural Science* (New Haven, CT: Yale University Press, 1975); Kloppenburg, *First the Seed*, 58; Paul Wallace Gates, *The Farmer's Age: Agriculture, 1815–1860: Economic History of the United States*, vol. 3 (New York: Holt, Rinehart and Winston, 1960); Fred A. Shannon, *The Farmer's Last Frontier: Agriculture, 1860–1897* (Armonk, NY: M. E. Sharpe, 1989).

2. Kloppenburg, *First the Seed*; Marina Moskowitz, "Broadcasting Seeds on the American Landscape," in *Cultures of Commerce: Representation and American Business Culture, 1877–1960*, ed. E. H. Brown, C. Gudis, and M. Moskowitz (New York: Palgrave Macmillan, 2006); Daniel Kevles, "Fruit Nationalism: Horticulture in the United States from the Revolution to the First Centennial," in *Aurora Borealis: Studies in the History of Science and Ideas in Honor of Tore Frängsmyr*, ed. M. Beretta, K. Grandin, and S. Lindquist (Sagamore Beach, MA: Science History Publications, 2008).

3. Pauly, *Fruits and Plains*, 99–130; Moskowitz, "Broadcasting Seeds on the American Landscape"; Marina Moskowitz, "The Limits of Globalization? The Horticultural Trades in Postbellum America," in *Food and Globalization: Consumption, Markets, and Politics in the Modern World*, ed. A. Nützenadel and F. Trentmann (Oxford: Berg, 2008), 57–73.

4. Crosby, *Columbian Exchange*; Crosby, *Ecological Imperialism*; William Cronon, *Changes in the Land: Indians, Colonists, and the Ecology of New England* (New York: Hill and Wang, 1983).

5. Kloppenburg, *First the Seed*, 50–65; Nelson Klose, *America's Crop Heritage* (Ames: Iowa State College Press, 1950), chaps. 3–4; Drew A. Swanson, *A Golden Weed: Tobacco and Environment in the Piedmont South* (New Haven, CT: Yale University Press, 2014).

6. Drawing on Peter H. Wood's thesis that South Carolina rice planters chose slaves from West Africa's rice-growing regions because of their expertise in cultivation, Judith Carney argues that West African slaves were the primary introducers of seed and knowledge to North America. Peter H. Wood, *Black Majority: Negroes in Colonial South Carolina from 1670 through the Stono Rebellion* (New York: Knopf, 1974), chap. 2; Judith Ann Carney, *Black Rice: The African Origins of Rice Cultivation in the Americas* (Cambridge, MA: Harvard University Press, 2001). Revisions and challenges to Carney's thesis are summarized by S. Max Edelson in "Beyond 'Black Rice': Reconstructing Material and Cultural Contexts for Early Plantation Agriculture,"

American Historical Review 115, no. 1 (2010): 125. A recent global history of rice cultivation and technology transfer is Francesca Bray et al., *Rice: Global Networks and New Histories* (New York: Cambridge University Press, 2015).

7. On Jefferson's efforts and the Hessian fly invasion, Pauly, *Fruits and Plains*, 9–50; Kloppenberg, *First the Seed*, 52; Klose, *America's Crop Heritage*, 15.

8. On cotton from Anguila, Sven Beckert, *Empire of Cotton: A Global History* (New York: Knopf, 2014), 100–101; Pauly, *Fruits and Plains*, 102. On experimentation with silk, indigo, cotton, and rice, Joyce E. Chaplin and Institute of Early American History and Culture (Williamsburg, Virginia), *An Anxious Pursuit: Agricultural Innovation and Modernity in the Lower South, 1730–1815* (Chapel Hill: Published for the Institute of Early American History and Culture, Williamsburg, Virginia, by the University of North Carolina Press, 1993).

9. On Crawford, Kloppenburg, *First the Seed*, 53–54; Klose, *America's Crop Heritage*, 13–26. On Richard Rush's request to consuls and the navy, Pauly, *Fruits and Plains*, 101; Richard Rathbun, "The Columbian Institute for the Promotion of Arts and Sciences," USNM Bulletin 101 (Washington, DC: Government Printing Office, 1917), 48.

10. On Ellsworth and Jacksonian rationalization of monopoly in invention, Pauly, *Fruits and Plains*, 104–5.

11. On the chronology of plant patents and the earlier exclusion of seeds as products of nature, see Daniel Kevles, "Patents, Protections, and Privileges: The Establishment of Intellectual Property in Animals and Plants," *Isis* 98 (2007): 323–31, and "A History of Patenting Life in the United States with Comparative Attention to Europe and Canada: A Report to the European Group on Ethics in Science and New Technologies" (Luxembourg: Office for Official Publications of the European Commission, 2002).

12. Richard R. John, *Network Nation* (Cambridge, MA: Harvard University Press, 2010), 433–35; E. Merton Coulter, *Daniel Lee, Agriculturist: His Life North and South* (Athens: University of Georgia Press, 2010), 31.

13. Pauly, *Fruits and Plains*, 104–5; John, *Network Nation*, 433–35.

14. S.Doc 95, 25th Congress, 3rd Session, January 28, 1839. "Letter from the Commissioner of Patents to the Chairman of the Committee on Patents and the Patent Office, in relation to the collection and distribution of seeds and plants."

15. Rathbun, "The Columbian Institute for the Promotion of Arts and Sciences," 37–54. The funds for the construction of the Patent Office came out of the $20,000 naval appropriation for the transit and care of the expedition's collections. Brackenridge continued to manage the garden until he was succeeded by William R. Smith sometime in the 1850s. Alfred Hunter indicates in the preface of his catalog of the exhibitions in the Patent Office Building that the catalog of the National Conservatory and its botanic specimens was prepared "by the skillful botanist in whose charge they flourish." William R. Smith was superintending botanist at the time of publication. Hunter, *A Popular Catalogue of the Extraordinary Curiosities in the National Institute, Arranged in the Building Belonging to the Patent Office*, 1857.

16. On the noncommodification of seeds in the context of recent debates over intellectual property rights for biological material, Keith Aoki, *Seed Wars: Controversies and Cases on Plant Genetic Resources and Intellectual Property* (Durham, NC: Carolina Academic Press, 2008).

17. Ellsworth's successor, Edmund Burke, had little interest in agriculture, producing an initial annual report that the agriculturalist Solon Robinson christened a "bundle of trash." He also thought the bulk of the seeds distributed entirely useless. The editor of the *Southern Cultivator* and *Genesee Farmer*, Daniel Lee, lamented the poor quality of the agricultural statistics in the report, compiled by means of a crops circular so general and vague Lee supposed

the authors knew nothing at all about agriculture. Burke's showing was so poor he managed to lose the agricultural appropriation for a year in 1846, with the House Committee describing the Agricultural Department as a dismal affair: a single clerk "pent up in the cellar of the Patent Office" with a candle at midday, "struggling to get up the report." Lee's agitations against Burke and his successor, Thomas Ewbank, inspired the new secretary of the interior to appoint him to the position of agricultural clerk. Lee revived the Agricultural Department, devoting special attention to the compilations of agricultural statistics on the southern states, and somewhat less to the importation and distribution of seeds. Nevertheless, Lee maintained a strong interest in diversifying southern agriculture, devoting special attention to the possibilities suggested by viniculture, sorghum, and pasture grasses. Coulter, *Daniel Lee, Agriculturist*, 31–34, 67–74.

18. D. J. Browne, "Preparation for a Government Propagating Garden at Washington," *Report of the Commissioner of Patents: Agriculture*, 1859, 280–82; "Government Experimental and Propagating Garden," *Report of the Commissioner of Patents: Agriculture*, 1859, 1–22; Kloppenburg, *First the Seed*, 55; Klose, *America's Crop Heritage*, 33; Pauly, *Fruits and Plains*, 107; Louis Joseph Brown, "The United States Patent Office and the Promotion of Southern Agriculture, 1850–1860" (master's thesis, Florida State University, Tallahassee, 1957).

19. Charles Stuart Kennedy, *The American Consul: A History of the United States Consular Service, 1776–1914*, Contributions in American History (New York: Greenwood Press, 1990).

20. The Patent Office requested that the Chilean consul prepare information on the country's agriculture. He formulated a plan to send tables of desired plants to the rector of the university for distribution and to an acquaintance who was an accomplished practical farmer in central Chile. Subsequently, the Patent Office established a seed exchange program with the Chilean government. J. M. Gillis, Washington, and January 6, 1854, replying to a letter of January 5 asking him to obtain information re: agricultural crops of Chile. RG 16.1, V.97, National Archives and Records Administration (hereafter NARA), Washington, DC; James H. Causten, Consulate of Chile, May 10, 1854, enclosing copy of instructions he received from the Secretary of State of Chile, NARA RG 16.1 V.577, 579. The consul in British Guiana was also especially active in seeking and forwarding specimens, including hemp grass, plantain fiber, India rubber, gums, tobacco, kakarali bark, sarsaparilla, and asphaltum. Charles W. Dennison, Consul, Georgetown, Demerara, British Guiana, March 18, 1854, requesting sanction for an expedition into the interior, regarding specimens and findings (April 30, 1854), and forwarding several packages of tropical productions (June 6, 1854), NARA RG 16.1, V.405, 529, and 613, respectively. The US Commercial Agency in Saint Helena also submitted vegetables, flowers, and fruits. George W. Kimball, US Commercial Agency, Saint Helena, March 1, 1854, NARA RG 16.1, V. 345.

21. J. C. Dobbin, Navy Department, May 3, 1854, New York Navy Yard sending botanical specimens collected by Lt. Thos. J Page in the progress of a reconnaissance of the river Paraguay. J. C. Dobbin, November 28, 1854, submitting a box of seeds, NARA RG 16.1, V. 555, 1239.

22. D. O. Williams, Clinton, MO, December 3, 1839, requesting information on cinchona. NARA RG 16.1, I.912.

23. Morrow and Cole, *Scientist with Perry*; C. Dobbin, Navy Department, January 23, 1855, on US Ship Plymouth from East Indies and Indian Ocean with Mauritius sugar cane seedlings and Cape Town wheat, NARA RG 16.1 IX.

24. Letter of October 9, 1854, Franklin Pierce and Matthew Calbraith Perry, Message of the President of the United States, Transmitting a Report of the Secretary of the Navy, in Compliance with a Resolution of the Senate of December 6, 1854, Calling for Correspondence, &c., Relative to the Naval Expedition to Japan ([Washington, DC]: [publisher not identified], 1855), 183.

25. NARA RG 16.1, XII 531. The vessel *Release* was specially prepared for the voyage, with a carpenter contracted to outfit the ship with proper fittings for plant and seed collection. XII 599, 601, XV 399. Ultimately, the collection was aided by Townend Glover, an entomologist who devoted significant research to preserving southern fruit trees from infestation and later became chief entomologist of the USDA: NARA RG 16.1, XII 983.

26. William Leach Giro, Consulate at Alicante shipped 8 barrels vica wheat, 5 barrels Alicante candela white wheat, 5 barrels Alicante guja red wheat, along with barley, lupines, maize, corn, rye, oats, chickpeas, hemp, flax seed, chufa bulbs, melon, carob, onion, olive, pepper, fenugreek, and almond: January 6, 1855, and March 5, 1855, NARA RG 16.1, VI. 529, 1086; J. Mahoney, US Consulate, Algiers, prepared twelve casks of the best quality flint wheat response to the July 1853 circular requesting seeds: October 19, 1854, NARA RG 16.1, V.921. Danzig firm recommended for spring wheat in preference to those suggested by Patent Office on account of shipping disruptions: NARA RG 16.1, VI. 535.

27. G. Neveu, Fond du Lac, Wisconsin, October 31, 1854, Chrystal wheat warranted special mention. De Neveau responded that he found it not worthy of general cultivation. Instead he sent three barrels (1.5 bushels) of Rio Grande Spring and Canada club wheat. NARA RG 16.1, V.987. Evidently, the Patent Office failed to pay the bill for the shipping. NARA RG 16.1, VI.832.

28. I. W. Buchanan, Tullahoma, Tennessee, to Commissioner of Patents, March 30, 1858, NARA RG 16.1, XX.54.

29. In August 1857, the Cuyahoga County Agricultural Society in Cleveland, Ohio, acknowledged 22 bags of red wheat from Turkey, as did the Delaware County Institute of Science of Media, Pennsylvania, the Vanderburg County Agricultural and Horticultural Society of Evansville, Indiana, and the Pittsburgh Horticultural Society of Pittsburgh, Pennsylvania, which also reported on Tuscan wheat. NARA RG 16.1, XVI. 465, 455, 504, 473.

30. In 1852, G. Schnabel of Williamsport, Pennsylvania, requested the Iona wheat collected by the Patent Office. NARA RG 16.1, XI. 987. A number of letters pertain to experimentation with Turkish flint wheat. On October 19, 1855, Hugh Craig of Jefferson, South Carolina, reported on Algerian wheat, red Spanish spring wheat, and Turkish flint wheat, claiming that the public mind was excited on the subject of small grain crops and reporting the formation of an agricultural society in his "fine wheat growing region of country," NARA RG 16.1, VII.881. In 1857, D. C. Ellsworth of Pittsfield, Ohio, requested Chilean and Algerian wheat and reported on Chinese sugarcane, NARA RG 16.1, XVI 470. On August 29, 1860, Peter Gormary, Laurel, Maryland, acknowledging 26 quarts of fall wheat from Syria, stating that from a half bushel of seed received from the Patent Office two years ago he now has 136 bushels of "the best wheat in the country." NARA RG 16.1, XXI 202.

31. On October 10, 1852, Isaac Newton of the Philadelphia Agricultural Society to Commissioner of Patents, requesting half a bushel of Mediterranean and Chilean wheat: NARA RG 16.1, II.1124.

32. July 1855, OH Kelly, Cor. Sec., Fenton Co Ag Soc of Northwood, MN to Commissioner of Patents, NARA RG 16.1, VII 599.

33. July 31, 1856, John Henry to Commissioner of Patents, NARA RG 16.1, XI.997.

34. Daniel Somers, Ravenswood, Virginia, March 17, 1854, submitting a package of seed, NARA RG 16.1, V.403.

35. John Reed, Huntington, Pennsylvania, sending bean seeds, Egyptian Mummy wheat, Irish Bartley, January 30, 1855, NARA RG 16.1, VI.840.

36. John H. Klippart, *The Wheat Plant, with Remarks on Indian Corn* (New York: A. O. Moore, 1860), 61, 514.

37. P. M. Lasher, postmaster, Tivoli, to Horatio King, 1st assistant postmaster general, Washington, July 6, 1854, NARA RG 16.1, V.873.

38. W. Noland to Burke, Aldie, September 13, 1848, submitting small sample of wheat found in the straw packed with Washington's statue, collected and grown by his grandson, Edmund Berkeley Esq. of Evergreen, NARA RG 16.1, I.107.

39. Warren & Co, Sacramento City, July 26, 1852, sending specimens from the residence of Captain John A. Sutter, NARA RG 16.1, II. 1044.

40. David I. Spanagel, *DeWitt Clinton and Amos Eaton: Geology and Power in Early New York* (Baltimore: Johns Hopkins University Press, 2014), 53–54. On British imperial scientific interest in wild rice as a pioneer plant, see Anya Zilberstein, "Inured to Empire: Wild Rice and Climate Change," *William and Mary Quarterly* 72, no. 1 (2015): 127–58.

41. Charles Mason, "Experiments with Seeds," US Patent Office Report for the Year 1854, Agriculture.

42. John P. Brown, Legation of US, Constantinople, July 16, 1854. Brown was also asked to forward a list of the chief garden, orchard, and field products of western Asia. NARA RG 16.1, V. 679.

43. John P. Brown, Constantinople, October 18, 1854, stating that he had shipped 27 bushels of wheat to Boston in the Barque Mystery. He also recommended that Mason investigate Tajan Rock [?] and Ishmael wheat, considered the "next best to Polish wheat," NARA RG 16.1, V.913.

44. Daniel Lee to Commissioner, NARA RG 16.1, XVI.56, 65. In 1849, the Patent Office, formerly managed by the Department of State, was transferred to the Department of the Interior. President Taylor created the new department ostensibly to manage the new western territories, but also in an attempt to balance the interests of northern and southern Whigs through appointments. On sectional conflict in the Patent Office, see Norman O. Forness, "The Seward-Fillmore Feud and the US Patent Office," *Historian* 54 (1992): 255–68.

45. "Preliminary Remarks," Report of the Commissioner of Patents for the year 1853. Agriculture, Serial Set Vol. No. 697, Session Vol. No.7, 33rd Congress, 1st Session, S.Exec.Doc. 27 pt. 2, p. v.

46. Beginning in 1856, the Patent Office also sponsored the collection and reduction of meteorological data, occasioning Joseph Henry's contribution to the agricultural reports on the connection of meteorology to agriculture. NARA, Records of the Agricultural Division of the Patent Office, 1839–60, United States Department of Agriculture, RG 16.1, correspondence between Joseph Henry and Commissioner of Patents, V.539, 643; VI.1090; X.449; XIV.625, 923; XVI.445.

47. Commissioner Edmund Burke acknowledges Vattemare's efforts in "Annual Report of the Commissioner of Patents" for 1847. Vattemare suggested that the commissioner "form in the Patent Office a kind of depot for all the most important seeds of North America for international exchanges. Alex. Vattemare of Paris to Commissioner of Patents, April 15, 1858, NARA RG 16.1, XXI.71; Alex. Vattemare to Commissioner, June 8, 1858, addressing in the name of the Field Marshall A. Vaillant, Minister of War, a small series of seeds proceeding from the Algerian Annual Agricultural Exhibition, and from the ministers own private gardens. NARA RG 16.1, Unnumbered [XXI.97].

48. C. P. Hagedorn, Consul for Bavaria, Baden, etc., Philadelphia, April 14, 1856, forwarding a few choice vegetable and flower seeds imported from Baden, and stating that the government of Baden is shipping seeds for the Patent Office. Notation: thank him for seeds and say to him that all the varieties have been purchased from the same house from which they came with many others within the last two years. If he sends any more seeds let them be procured

direct from Bavaria, NARA RG 16.1, XI.301; C. P. Hagedorn, Bavarian Consulate, Philadelphia, November 19, 1856, stating that he has been directed to report on the culture of tobacco to the government of Bavaria, and to send tobacco seed, and requesting assistance. Notation: refer him to B. L. Jackson & Bros., Pa. Society for Seed. NARA RG 16.1, XII.753.

49. Vilmorin Andrieux and Co., Paris, May 27, replying to a letter of March 29, stating it was not at their suggestion that Alexander Vattemare requested an acknowledgment of the receipt of the seeds offered to the Patent Office through him, that the firm is glad to learn that the seeds are placed in the collection of the museum, NARA RG 16.1, XXI.98. When the Patent Office's seed programs were investigated in 1858, William R. Smith testified to the quality of the seed received from both Vilmorin and Charlwood & Cummings. W. R. Smith to Commissioner Holt, Botanic Garden, Washington, June 14, 1858, NARA RG 16.1, XX.98.

50. Vilmorin, Andrieux, Paris, December 31, 1860, re: shipment of 128 cases of grains, NARA RG 16.1, XX1.264.

51. Wm R. Prince, Prince's Linnaean Botanic Garden and Nurseries, Flushing, New York, November 4, 1858, re: Secretary of the Interior's views with regard to establishing a national nursery. Approves, after the French model.

52. William G. Whitely, Chairman, Committee on Agriculture, Report of the Committee, January 10, 1859, NARA RG 16.1, XXI.8. Also see XX.95, 96 regarding initial investigation.

53. Samuel S. Cox, Columbus, OH, May 12, 1857, NARA RG 16.1, XVI.267; T. D. Jones, Winterset, January 3, 1858, requesting Patent Office reports, NARA RG 16.1, XX 4.

54. Edmund Ruffin, *The Diary of Edmund Ruffin: Toward Independence, October 1856–April 1861* (Baton Rouge: Louisiana State University Press, 1972), 263–65. On Ruffin's career as an agriculturalist, see Steven Stoll, *Larding the Lean Earth: Soil and Society in Nineteenth-Century America* (New York: Hill and Wang, 2002), 143–96.

55. William G. Whitely, Chairman, Committee on Agriculture, Report of the Committee, January 10, 1859, NARA RG 16.1, XXI.8. Also see XX.95, 96 regarding initial investigation.

56. H. C. Williams, near Jefferson, Texas, to Gregory A. Alexander, August 12, 1860, NARA RG 16.1, XXI.196.

57. Isaac Newton, President, Pennsylvania Agricultural Society, Philadelphia, PA, October 10, 1852, recommending David Landreth as being "one of the best practical seedsmen we have in the United States to select foreign and home seeds, NARA RG 16.1, II.1124, 1184, 1186; Landreth to Commissioner, NARA RG 16.1, III.1115.

58. Critical remarks of T. T., Philadelphia, on Landreth, July 1, 1858, quoting the Pa. Hort. Soc. Rept., May 18, 1858, and the North American Newspaper, June 26, 1858, NARA RG 16.1, XX.85.

59. Pennsylvania Horticultural Society, Philadelphia, May 18, 1858 to Holt, protesting distribution of seed. NARA RG 16.1, XX.86. Pages missing.

60. P. A. Rett, Pres. Southern Agricultural Society of Louisiana, Washington, September 4, 1857, to John Jacob Thompson, Secretary of the Interior, NARA RG 16.1, XVII.14.

61. In 1856, J. R. Thomas's circular was reprinted and adapted for the *New York Observer*, the *Washington Union*, the *National Intelligencer*, the *German Reformed Messenger*, and numerous agricultural periodicals, including the *Prairie Farmer*, *Ohio Farmer*, *Michigan Farmer*, *American Farmer*, *Cultivator*, and others.

62. A general ethnography is Georges E. Sioui, *Huron-Wendat: The Heritage of the Circle*, trans. Jane Brierley (Vancouver: University of British Columbia Press, 1999). The definitive account of Huron history is Bruce Trigger, *Children of Aataentsic: A History of the Huron People to 1660* (Montreal: McGill-Queen's University Press, 1976). On the agricultural history of the

Huron, see Bruce Trigger, *The Huron: Farmers of the North* (New York: Holt, Rinehart and Winston, 1969).

63. Advertisements appeared regularly in the *Charleston Mercury* and the *Augusta Chronicle* between 1856 and 1858. *Southern Cultivator* also had a number of articles on the corn.

64. G. C. Worth, *Ohio Farmer* 5 (February 1856).

65. The *Friends' Intelligencer* reported it "not more valuable than other varieties in general use," including King Phillip, Dutton, Baden, Lee, and Oregon corns. *Friends Intelligencer* 12 (February 2, 1856): 735; *Southern Planter* 18 (February 1858): reprint of earlier piece from the *Farm Journal*, signed WLR Feb 1856.

66. *Ohio Farmer* 5 (February 1856); *Michigan Farmer* 16 (December 1859).

67. *Southern Planter* 18 (February 1858): reprint of earlier piece from the *Farm Journal*, signed WLR Feb 1856.

68. *Friends' Intelligencer* 12 (Wm. W. Moore, 1856); John Henry, Mount Erin, Indiana, July 21, 1856, requested Wyandotte corn and reported favorably on Turkish white flint wheat. NARA RG 16.1, XI 997; R. Buist, Philadelphia, April 8, 1857, stating that he is sold out on Wyandot corn, NARA RG 16.1, XVI.153.

69. Which one writer dubbed "profligate corn," *Southern Planter* 18 (April 1858).

70. Correspondence regarding Wyandotte and other commercial varieties persisted in the columns of *American Farmer*, which ran the circular through the late 1850s: *American Farmer*, printed for John S. Skinner, by J. Robinson, 1866.

71. Edward Stabler of Hareword, MD, May 2, 1856, sent 1/2 bushel of Wyandot corn for distribution, "having now supplied all applicants so far, even to South Carolina and Alabama, through the published notice in the Washington paper." NARA RG 16.1, XI.461; J. C. Thompson of Staten Island, New York, July 28, 1856, sent a letter inquiring about the results of planting Wyandot corn and enclosing a circular advertising it. The Office replied that it had received many complaints that the Wyandot corn did not vegetate. NARA RG 16.1, XI.979, 981. Not dissuaded, Thompson sent another circular on the Wyandot corn on November 23, 1856, XII.767. One farmer in Mount Erin, Indiana, indicated that the Turkish white flint wheat sent by the Patent Office had produced a good crop, also requesting Wyandot corn for a trial. John Henry of Mount Erin, Indiana, to Commissioner of Patents, July 31, 1856, NARA RG 16.1, XI.997.

72. Klippart, *Wheat Plant*, 666. Klippart also reviewed a number of other proprietary breeds, including Peabody's Prolific, which he noted was considered by some to be a humbug. He withheld judgment, citing Dr. Warder's description of its physiognomy and resemblance to another popular variety, Early Adams.

73. As an open-pollinating species, maize readily crossed with other varieties. In 1883, the New York State Agricultural Society downplayed differences between subspecies of corn, noting that few plants were as "subject to wider variation of habit" within a single variety: *Transactions of the New-York State Agricultural Society for the Year 1883* (New York: The Society, 1883); *Annual Report of the Board of Control of the New York Agricultural Experiment Station*, nos. 4–5 (Geneva, Ontario County: State Printers, 1886).

74. Chester Root to Commissioner Edmund Burke, Mobile, AL, January 8, 1848, submitting a sample of bear grass, used for making rope by Indians, NARA RG 16.1, I.83; Philander Prescott, Superintendent of Farming for the Sioux, November 10, 1849, NARA RG 16.1, I.730.

75. Kloppenburg, *First the Seed*, 65–90; Shannon, *Farmer's Last Frontier, Agriculture, 1860–1897*.

76. Pauly, *Fruits and Plains*, 112–30.

77. Kloppenburg, *First the Seed*, 91–129.

CHAPTER THREE

1. "Dead Letters—by a Resurrectionist: Written for the Albany Register," *Albany Register*, September 23, 1852 (misdated September 22).

2. *Ladies' Companion and Literary Expositor*, 1843, "The Culture and Preparation of Tea."

3. Peter Perdue, "Is Pu-Er in Zomia? Tea Cultivation and the State in China," Agrarian Studies Colloquium, Yale University, New Haven, CT, 2008.

4. Robert Paul Gardella, *Harvesting Mountains: Fujian and the China Tea Trade, 1757–1937* (Berkeley: University of California Press, 1994).

5. David Arnold, *Science, Technology, and Medicine in Colonial India* (Cambridge: Cambridge University Press, 2000), 47–55; Lucille Brockaway, *Science and Colonial Expansion: The Role of the British Royal Botanic Gardens* (New York: Academic Press, 1979), 26–29.

6. Perhaps the first of these appeared in the *American Museum, Repository of Ancient and Modern Fugitive Pieces* (Philadelphia) in August of 1787, p. 71. A brief column likens China's geography to the United States in terms of ocean exposure, surrounding territory, latitude, and native plants. Subsequent advocates of tea cultivation called on the same basic arguments. Another early proponent was the horticulturalist and landscape designer Andrew Jackson Downing in the *Farmer and Gardener and Livestock Breeder and Manager*, May 10, 1836. "The Tea Plant, Its Culture in the US," appeared in the September 1848 issue of *Plough, Loom, and Anvil*, an influential monthly published by agriculturalist B. F. Skinner. It was reprinted frequently in *Scientific American* (October 7, 1848) and other popular technical, agricultural, and literary periodicals. Proclamations regarding climactic similarities were echoed in the correspondence of the Patent Office Agricultural Department, e.g., vol. 11, no. 157, an Alabama planter remarking on similarities between East Asia and the southern United States.

7. On the silk prospecting boom of the 1820s and the Patent Office's crop introduction efforts, see Pauly, *Fruits and Plains*, 102–4.

8. In 1847, two years after the end of his first voyage, Fortune published his first book, *Three Years' Wanderings in the Northern Provinces of China* (London: J. Murray, 1847). Although Fortune collected no tea on this trip, his narrative featured "a Visit to the Tea, Silk, and Cotton Countries: With an Account of the Agriculture and Horticulture of the Chinese, New Plants, etc.," and this section attracted the most interest. Fortune also recounted his observations in his 1852 and 1853 publications, *A Journey to the Tea Countries of China* and *Two Visits to the Tea Countries of China*, both widely reviewed in the American literary, horticultural, and agricultural press.

9. For example, on yellow camellias, Fortune, *Three Years' Wanderings*, 82, and on restricting visits to nurseries, 119–24.

10. Fortune has a somewhat different interpretation of the registration policy, characterizing Chinese tradesmen as skittish and their governors as corrupt. See *Three Years' Wanderings*, 8.

11. Jayeeta Sharma, "'Lazy' Natives, Coolie Labour, and the Assam Tea Industry," *Modern Asian Studies* 43, no. 6 (2009): 1287–324.

12. On Chinese worker rebellions and the Singpho uprising at Sadiya in 1843, see Jayeeta Sharma, "British Science, Chinese Skill, and Assam Tea," *Indian Economic Social History Review* 43 no. 4 (2005): 429–55, and Sharma, "'Lazy' Natives, Coolie Labour, and the Assam Tea Industry."

13. Sharma, "'Lazy' Natives, Coolie Labour, and the Assam Tea Industry," 73, 77.

14. Francis Bonynge, *The Future Wealth of America, Being a Glance at the Resources of the United States and the Commercial and Agricultural Advantages of Cultivating Tea, Coffee, and*

Indigo, the Date, Mango, Jack, Leechee, Guava, and Orange Trees, Etc., with a Review of the China Trade, by Francis Bonynge, for Fourteen Years a Resident in India and West of China (New York: Published by the author, 1852), 86.

15. Fortune, *Three Years' Wanderings*, 190–227.

16. Perdue, "Is Pu-Er in Zomia?"; Sidney Wilfred Mintz, *Sweetness and Power: The Place of Sugar in Modern History* (New York: Viking, 1985).

17. *Niles Weekly Register*, Baltimore, August 28, 1817, Agricultural Index, II.215. "Hyson Tea," Fayetteville, NC, *Niles Weekly Register*, June 7, 1823. "Brief Notes on the Ag Resources of S Carolina," in the *Southern Agriculturalist and Register of Rural Affairs*, January 1828, p. 17. For a general summary, see Klose, *America's Crop Heritage*.

18. Petition of Newbold N. Puckett & Co., for a patent for the cultivation of the tea plant. March 17, 1846, Submitted by Senator Mangrum of North Carolina. Referred to the Committee on Patents and the Patent Office, and ordered to be printed. S.Doc. 227, 29th Congress, 1st Session. "Newbold N. Puckett" appears alternately as "Newbald M. Prickett" and "Newbald N. Prichett" in the *Congressional Globe* and the *Journal of the Senate*, respectively. The memorial probably met its end at committee. No patent appears to have been granted, and the Senate Committee records for the 29th Congress have been lost. (Manuscript records of the Committee on Patents and the Patent Office survive for all other Congresses.)

19. For example, some methods of manuring (e.g., Bommer's method) received patents, albeit seldom given the likelihood of collecting licensing fees or stopping piracy/replication. In colonial America, gristmills and sawmills, for example, received patents. In this sense, early patents bore some similarity to charters of incorporation. On incorporation, see Morton Horwitz, *The Transformation of American Law, 1780–1860* (Cambridge, MA: Harvard University Press, 1979); Gary Kulik, "Dams, Fish, and Farmers: Defense of Public Rights in Eighteenth-Century Rhode Island," in *The Countryside in the Age of Capitalist Transformation: Essays in the Social History of Rural America*, ed. Steven Hahn and Jonathan Prude (Chapel Hill: University of North Carolina Press, 1985). Puckett's petition also had precedents in Congress's tentative efforts to support crop introduction by funding private experiments in grapes and olives. See, for example, Klose, *America's Crop Heritage*, 24–25.

20. Junius Smith, Letter to Commissioner of Patents, December 11, 1849, NARA RG 16.1.

21. The same year, the missionary and diplomat Samuel Wells Williams had published his first edition of *The Middle Kingdom*, billed as a survey of the geography, government, literature, social life, arts, and history of the Chinese empire and its inhabitants (New York: Wiley and Putnam, 1848), with substantial notes on tea cultivation based on his earlier publications in the *Chinese Repository*. The *Chinese Repository*, which ran reports critical of Fortune's lack of knowledge, was a monthly magazine published by E. H. Bridgeman between 1832 and 1851, of which Williams eventually became editor.

22. Letter to Henry Smith, August 3, 1848. All letters between Junius and Henry Smith reprinted in Leroy Pond, *Junius Smith: Biography of the Father of the Atlantic Liner* (1927; Manchester, NH: Ayer Publishing, 1971).

23. Letter to Henry Smith, July 6, 1848. His pamphlet, "On the Cultivation of the Tea Plant in the United States," was published in the Annual Report of the American Institute of the City of New York in March 1848, directed by Smith's friend Henry Meigs (pages 271–310). Meigs gave the institute frequent reports of Smith's progress and advocated the cause of tea cultivation after Smith's death. Smith also self-published the pamphlet to gain publicity for his cause, sending instructions to his nephew to distribute them in fourteen states, principally in southern climates but also in centers of business in Philadelphia and New York.

24. Later Smith cited his many experiments in conveyance as part of his hard-earned expertise in introducing tea. But Fortune and other collectors' use of Ward cases made Smith's claims seem unpersuasive. NARA RG 16.1, Letter to Commissioner of Patents, December 11, 1849 (also printed in Annual Report of following year).

25. A list of subscribers published in *The Future Wealth of America* lists twenty-nine subscribers, including merchants, planters, and the South Carolina Agricultural Society. Bonynge, *Future Wealth of America.*

26. Ibid., 67.

27. Ibid., 86.

28. *Southern Cultivator,* November 1850, 68.

29. *Journal of the Senate,* July 20, 1852. Mr. Rusk presented the memorial of Francis Bonynge, praying for an appropriation to enable him to introduce and cultivate the tea, indigo, coffee, and other tropical plants and fruits into the United States, which was referred to the Committee on Agriculture.

30. *Scientific American,* July 1, 1854, reprint from *Dunkirk Journal.*

31. Alfred Hunter, *A Popular Catalogue of the Extraordinary Curiosities in the National Institute, Arranged in the Building Belonging to the Patent Office. Curiosities collected from all parts of the world, by the officers of the Army and Navy of the United States* . . . 2nd ed. (Washington, DC: 1857). Hunter's catalog is discussed at length elsewhere in the manuscript; he also published a catalog of the adjoining halls of patent models.

32. While westerners commonly held that green and black teas came from two different plants, Fortune verified accounts of several English observers that the varieties were distinguished only by their mode of processing (*A Journey to the Tea Countries of China* [London: J. Murray, 1852], and *Two Visits to the Tea Countries of China* [London: J. Murray, 1853], 187 ff.). Majorbanks's report to the Parliament had observed the common botanical source of the two teas as early as 1830, but controversy on this point persisted until Fortune's publication. Westerners also debated the likelihood of adulteration, many claiming that green teas were green because they were colored with Prussian blue or other harmful dyes, and others reporting that foreign and inferior leaves were mixed with finer teas for the undiscriminating English market.

33. Samuel Wells Williams was also instrumental in the negotiation of the 1857 treaty of Tianjin in his capacity as secretary-interpreter of the US legation to China, relaxing restrictions on US trade with China. This and other unequal treaties signed by the British, French, and Russians concluded the Second Opium War between the Qing Dynasty and British and French governments.

34. The records of the Patent Office Agricultural Division acknowledge correspondence regarding plants of China in the year 1849 from Davis, Macgowan, Bridgman, Balestiere, Huffnagle, Williams, and McCartee. McCartee and Macgowan did their best to supply Williams with seeds from regions around Shanghai and Ningbo, with McCartee noting that the lateness of the season reduced the stock and made locals unwilling to part with the remainder. Subsequent efforts were more successful: J. M. Davis to Commissioner of Patents, January 26, 1849; McCartee to Samuel Wells Williams, February 20, 1849; Williams to Davis, June 1, 1849; Williams to Davis, December 14, 1849; Davis to Commissioner Thomas Ewbank, December 17 and 27, 1849, all NARA 16.1.

35. McCartee's letters included extensive notes on fertilizers and cultivation. E. V. Bridgeman, the publisher of the *Chinese Repository,* to which Williams contributed and which Smith consulted in his research, forwarded pamphlets on tea, silk, and cotton to Burke's office, possibly also including samples of seeds: NARA RG 16.1. E. V. Bridgeman to Commissioner of

Patents, November 30, 1849. An attached page notes distribution of pamphlets by the Patent Office.

36. By 1846 Smith had already submitted reports on the cultivation of celery and broccoli. Also see his February 28, 1849, report to the New York Farmer's Club regarding seed potatoes, to which Smith also reported his success in securing and growing Chinese tea plants of seven years' growth in December 1848. S.Doc. 75, 28th Congress, 2nd Session; S.Doc. 307, 29th Congress, 1st Session.

37. Junius Smith to Commissioner of Patents, April 11, 1850, NARA RG 16.1, II.588.

38. The Patent Office Report of 1848 mentioned this news along with the success of contemporaneous tea experiments in Brazil, and the tentative progress of the British in Assam in producing tea for the British market: H.Exec.Doc. 59, 30th Congress, 2nd Session. A number of publications, including the *Hartford Daily Courant, Boston Medical and Surgical Journal,* and *Journal of Commerce and Art,* picked up the report.

39. Macgowan to Commissioner of Patents, letters dated February 1, February 30, and June 21, 1856, NARA RG 16.1.

40. India is "determined to distance China and supply herself and Europe with tea," Smith wrote. "How vain! That business is reserved for the United States and steam navigation." Letter to Henry Smith, August 1848, reprinted in Pond, *Junius Smith.*

41. "The Tea Plant, Its Culture in the US," *Plough, Loom, and Anvil,* September 1848. The article referred to a "resident returned from Calcutta, five to six years in management of one of the company's tea factories (i.e., plantations) in Assam" who had written a history of tea's culture and preparation. The *Friends Weekly Intelligencer* of January 26, 1850, provides the same description of a Philadelphia resident and names him as Spencer Bonsall. No patents for tea processing machinery appear to have been issued to Spencer Bonsall, but Bonsall did submit a report on Assam botany for the Patent Office reports and another on tea, possibly the one published in the 1860 report, NARA RG 16.1 XXI. 153, 170, 218.

42. *Merchant's Magazine and Commercial Review,* August 1859.

43. Bonynge's plantation was on land ceded to British control as an outcome of the Anglo-Burmese War of 1823–26. Francis Bonynge, *The Future Wealth of America* (University of California, [1852]), 208.

44. Coulter, *Daniel Lee, Agriculturist,* 68.

45. Report of the Commissioner of Patents for the year 1857. Agriculture, Serial Set Vol. No. 954 Session Vol. No. 8, 35th Congress, 1st Session, H.Exec.Doc. 32 pt. 4, 1858, preceding text of report.

46. Griffith also made this argument in his Report on the Tea Plant of Upper Assam, by Wm Griffith, Asst Surgeon Madras Establishment, late member of the Assam Deputation, Presented to Parliament as part of the 1839 Parliamentary Papers Relating to Measures for Introducing Cultivation of Tea Plant in British Possessions in India, Vol. XXXIX, Paper 63 (Calcutta, 1840). Griffith and Wallich's report was reviewed in the American press. For example, *American Journal of Science and Arts,* October–December 1840, p. 165.

47. Williams to Commissioner of Patents, May 21, 1851, NARA RG 16.1; H.Exec.Doc. 65 pt. 4, 34th Congress, 3rd Session and H.Exec.Doc. 32 pt. 4, 35th Congress, 1st Session. Browne answered the labor question with by now customary assertions: internal improvements in transportation would speed tea to market. He relied on Bonsall's speculations regarding labor-saving machinery. He also repeated the argument of Bonynge and Kew botanists that American labor was more reliable and robust than that of Asiatic bodies, drawing on Griffith's *Report on the Tea Plant of Upper Assam.* See Sharma, "British Science, Chinese Skill, and Assam Tea."

48. Fortune to Patent Office, November 19, 1857, NARA RG 16.1.

49. Commissioner of Patents to Robert Fortune, September 17, 1857, NARA RG 16.1; Fortune to Charlwood and Cummins, October 1, 1857. The records of the Agricultural Division contain Fortune's correspondence accompanying bills of lading for 1858 (August 10, December 6 and 14) and 1859 (January 22, February 19, May 24). Fortune responded angrily to his dismissal, demanding severance (August 5). Having received it, he agreed to supply answers to "interrogatories concerning the culture of the tea plant in China" forwarded to him by Holt. Fortune to Commissioner of Patents, September 6 and November 19, 1859, NARA RG 16.1.

50. *Southern Cultivator*, August 1859, p. 249. The seeds were distributed in the spring and summer of 1860. In July 1860, *The Valley Farmer* reported that plants were distributed to southern states, with congressional representatives determining intelligent and responsible recipients.

51. Hunter, *A Popular Catalogue of the Extraordinary Curiosities in the National Institute, Arranged in the Building Belonging to the Patent Office*, 64–71. The site of the propagating garden had been a topic of discussion before the site at First and Pennsylvania was selected. Curtis & Deaning, Washington, DC, October 12, 1858, offering proposals to erect a propagating house on the public square situated on Missouri Ave., between 4&1/2 and 6th St., NARA RG 16.1, XX 174.

52. Problems with the distribution of tea plants echoed those of the seed program more generally. *Southern Cultivator*, August 1859, Letter from S. D. on visiting Washington propagating garden; *Merchant's Magazine and Commercial Review*, August 1859. The records of the Agricultural Division (NARA RG 16.1) contain letters of receipt for tea seeds. John L. Phillips of Saint Augustine, Florida, to Commissioner of Patents, March 17, 1860, acknowledged receipt and distribution of 6 tea plants and enclosed 7 dollars for expressage. John L. Bridgers of Tarboro, North Carolina, to Commissioner of Patents, July 21, 1860, acknowledged receipt of four boxes of tea plants, and noted that he distributed a good number of them so that they might have a chance to grow in different soils but that by March they commenced dying, seemingly from disease. Louis Baker reported that contrary to (Unionist) Commissioner of Patents Thomas G. Clemson's charges, he did not steal tea seeds or sell them, but rather distributed them free of charge after having received them from his acquaintance (and Confederate) Gregory Alexander. He added that he needn't have bothered given that they were dead or dying. Louis Baker to Gregory Alexander, Washington, June 23, 1860, XXI.174.

53. Report of the Commissioner of Agriculture for the Year 1863, H. Ex. Doc. No. 91, 38th Congress, 1st Session, 514–517; Report of the Industrial Commission on Agriculture and Agricultural Labor, 1901, H. Doc. No. 179, 57th Congress, 1st Session, CCXXIX–CCXXX.

54. See Klose, *America's Crop Heritage*, 83.

55. "The Pinehurst Tea Farm: The Japanese Minister Visits Dr. Shepard's Plantation," *New York Times*, March 2, 1902, 15.

56. *Handbook of South Carolina Resources, Institutions and Industries of the State*, 2nd ed. (Columbia: South Carolina State Department of Agriculture, 1908), 324–27; Liberty Hyde Bailey, *Cyclopedia of American Horticulture* (London: Macmillan, 1902), 1773–75.

57. The *Wall Street Journal* is not sanguine about its success in a marketplace saturated with cheap offerings and boutique brands from exotic locales: "U.S. Tea Grower Is in Hot Water," *Wall Street Journal*, September 13, 2000. As of 2008, the plantation appears to be under the management of Bigelow Tea Company, purveyors of blended teas from various geographic sources.

58. Perdue, "Is Pu-Er in Zomia? Tea Cultivation and the State in China"; Gardella, *Harvesting Mountains*.

59. Mintz, *Sweetness and Power*.

60. Sharma, "'Lazy' Natives, Coolie Labour, and the Assam Tea Industry."

61. "Controversy over Assam's First Indigenous Tea Planter," *oneindia*, September 26, 2007, http://news.oneindia.in/2007/09/26/controversy-over-assams-first-indigenous-tea-planter-1190799863.html. This article draws on D. K. Taknet's *The Heritage of Indian Tea: The Past, the Present, and the Road Ahead* (Jaipur: IIME, 2002). Citing William Robinson's "Descriptive Account of Assam," objectors note that the British had already appointed King Nirula Singpho superintendent of tea growing plantations by its publication in 1841, whereas Maniran Dewan applied for rights to grow tea in 1853.

62. Sharma, "'Lazy' Natives, Coolie Labour, and the Assam Tea Industry."

FIELD NOTES: "LOCAL KNOWLEDGE"

1. Or mētis in James C. Scott's usage. See, for example, the working definition of "local knowledge" provided by the Economic and Social Development Department of the Food and Agriculture Organization (FAO) of the United Nations, and James C. Scott, *Seeing Like a State: How Certain Schemes to Improve the Human Condition Have Failed* (New Haven, CT: Yale University Press, 1998), 6–7.

2. On defiance, see James C. Scott, most recently: *The Art of Not Being Governed: An Anarchist History of Upland Southeast Asia* (New Haven, CT: Yale University Press, 2010). On pastoralism as a flexible continuum of economic strategies, see Anatoly Khazanov, *Nomads and the Outside World* (Madison: University of Wisconsin Press, 1994).

3. For this and other details on the entomology of Ararat cochineal, see Harald Böhmer, "Natural Dyes in the Near Middle East and India," in *Flowers Underfoot: Indian Carpets of the Mughal Era*, by Daniel S. Walker (New York: Metropolitan Museum of Art, 1997), 155–57, and R. Sarkisov, L. Mkrtchyan, V. Zakharyan, and A. Sevumyan, *Ararat Cochineal* (Yerevan, Armenia: Gitutyun, 2010).

4. Kelly Kindscher, *Medicinal Wild Plants of the Prairie: An Ethnobotanical Guide* (Lawrence: University Press of Kansas, 1992), 3–5; Donald Worster, *The Wealth of Nature: Environmental History and the Ecological Imagination* (New York: Oxford University Press, 1994), 45–70.

5. James Earl Sherow, *The Grasslands of the United States: An Environmental History* (Santa Barbara, CA: ABC-CLIO, 2007). On the increase in grass consumption for reconfigured food systems, Natale Zappia, "Revolutions in the Grass: Energy and Food Systems in Continental North America, 1763–1848," *Environmental History* 21, no. 1 (2016): 30–53.

6. Pekka Hämäläinen, "The Politics of Grass: European Expansion, Ecological Change, and Indigenous Power in the Southwest Borderlands," *William and Mary Quarterly* 67, no. 2 (2010): 173–208; Andrew C. Isenberg, *The Destruction of the Bison: An Environmental History, 1750–1920* (New York: Cambridge University Press, 2000), 22–23.

7. Frederick Jackson Turner, "The Significance of the Frontier in American History," in *The Frontier in American History*, by Frederick Jackson Turner (New York: Henry Holt, 1921), 293.

8. Theodore Roosevelt, *The Winning of the West* (New York: Putnam's, 1889); Richard Slotkin, "Nostalgia and Progress: Theodore Roosevelt's Myth of the Frontier," *American Quarterly* 33, no. 5 (1981): 608–37, doi:10.2307/2712805.

9. Judith Schwarz, "Yellow Clover: Katharine Lee Bates and Katharine Coman," *Frontiers: A Journal of Women's Studies* 4, no. 1 (1979): 59–67.

CHAPTER FOUR

1. Christian Krehbiel, *Prairie Pioneer: The Christian Krehbiel Story*, ed. Elva Krehbiel Leisy (Newton, KS: Faith and Life Press, 1961).

2. Three detailed accounts of Mennonite immigration to North America are Henry Smith, *The Story of the Mennonites, [by] C. Henry Smith* . . . (Berne, IN: Mennonite Book Concern, 1945); Cornelius Krahn et al., *From the Steppes to the Prairies, 1874–1949* (Newton, KS: Mennonite Publication Office, 1949); Royden Loewen, *Family, Church, and Market: A Mennonite Community in the Old and the New Worlds, 1850–1930* (Urbana: University of Illinois Press, 1993). On the varied impetus for migration, refer to note 27 in this chapter.

3. Quisenberry and Reitz, "Turkey Wheat." See also Robert G. Dunbar, "Turkey Wheat: A Comment," *Agricultural History* 48, no. 1 (1974): 111–14.

4. William Gaud coined the term in 1968 as a counterpoint to the red revolutions in the Asian countryside, arguing that the rural development activity of the last two decades would mean bumper harvests and happy farmers. Cullather, *Hungry World*, 7.

5. N. I. Vavilov, *World Resources of Cereals, Leguminous Seed Crops and Flax: And Their Utilization in Plant Breeding* (Moscow: Academy of Sciences of the USSR [Available from the Office of Technical Services, US Department of Commerce, Washington, DC], 1960), 7. This is a posthumous publication of Vavilov's research based on a stenogram left among his papers. Thanks to Josephine Piggin for this reference.

6. Mark Alfred Carleton, "Hard Wheats Winning Their Way," in *US Department of Agriculture, Yearbook* (Washington, DC: Government Printing Office, 1914), 407.

7. For example, the following influential accounts of wheat improvement all rely exclusively on Carleton's narrative: Carleton Roy Ball, "The History of American Wheat Improvement," *Agricultural History* 4, no. 2 (1930): 48–71; Jacob Allen Clark, Carleton Roy Ball, and John Holmes Martin, "Classification of American Wheat Varieties," US Dept. of Agriculture Bulletin No. 1074 (Washington, DC: Government Printing Office, 1922); S. C. Salmon, O. R. Mathews, and R. W. Leukel, "A Half Century of Wheat Improvement in the United States," *Advances in Agronomy*, vol. 5 (New York: Academic Press, 1953), 1–151; Moon, "In the Russians' Steppes." Numerous Mennonite heritage publications tout the introduction of Turkey Red wheat as a major achievement of pioneer settlers. One example is Michael L. Olsen, "And a Child Shall Lead Them: The Legendary Introduction of Turkey Red Wheat into Kansas," in *Great Mysteries of the West*, ed. Ferenc Morton Szasz (Golden, CO: Fulcrum, 1993).

8. John H. Klippart lists "Turkey large red" and "Caucasian red" as varieties under cultivation in *The Wheat Plant*, 496–98.

9. *The Cultivator*, new series, vol. 3, 1846. Numerous mentions of Black Sea wheat. On conversion to winter wheat, p. 303.

10. Pauly, *Fruits and Plains*, 49.

11. Ball, "The History of American Wheat Improvement"; G. M Paulsen and J. P. Shroyer, "The Early History of Wheat Improvement in the Great Plains," *Agronomy Journal* 100, no. 3 (2008): 70; Alan L. Olmstead and Paul W. Rhode, "The Red Queen and the Hard Reds: Productivity Growth in American Wheat, 1800–1940," *Journal of Economic History* 62, no. 4 (2002): 929–66; Olmstead and Rhode, *Creating Abundance*.

12. Olmstead and Rhode, *Creating Abundance*.

13. Paulsen and Shroyer, "Early History of Wheat Improvement in the Great Plains."

14. Carleton, "Hard Wheats Winning Their Way."

15. On the ecology of the steppe, see David Moon, *The Plough That Broke the Steppes: Agri-*

culture and Environment on Russia's Grasslands, 1700–1914, Oxford Studies in Modern European History (Oxford: Oxford University Press, 2013).

16. Carleton's trip to southern Russia in 1900 is documented in a series of letters with Mark Carleton, preserved in the National Archives and in Photostat copy at the Mennonite Archives at Bethel College (hereafter Warkentin Papers, Bethel College). National Archivist T. R. Schellenberg prepared a summary memorandum of Carleton's correspondence for Cornelius Krahn, which is preserved at in the Mennonite Archives at Bethel College: Summary of Warkentin correspondence with Mark Carleton, by T. R. Schellenberg, 1944, Box 1, Folder 18, Warkentin Papers, Bethel College, Newton, Kansas. A summary of Carleton's travels is provided by Moon, "In the Russians' Steppes."

17. Except where noted, the narrative of Warkentin's travels is gathered from his correspondence with David Goerz, Warkentin Collection, Bethel College, and his 1872 travel diary, preserved in the Warkentin Papers, State Archives of the Kansas Historical Society (hereafter KSHS), Topeka, Kansas. Warkentin's correspondence with childhood friend, emigrant, and business associate David Goerz provides a comprehensive account of his travels through the American West from 1872 to 1875.

18. Warkentin's letters about America circulated among congregations in the area, and the *Nordische Presse*, a Saint Petersburg paper, reported that three wealthy young Mennonites and a foreigner had traveled to the United States to negotiate with leaders there on behalf of their people. Warkentin's father, a wealthy and influential citizen, faced threats for encouraging resettlement. Other indications were no more promising. Delegations to Saint Petersburg had secured few concessions from Russian officials with respect to compulsory state service and Russian education. Goerz also reported of hostile encounters between Mennonites and Russian agricultural laborers in several of his early letters from Berdiansk.

19. Bernard Warkentin, Letter to Goerz, July 29/11, 1872, Warkentin Papers, Bethel College.

20. Krehbiel tells his story in an autobiography written in 1906 at the age of seventy-five. Except where noted, my account is drawn from the biography as translated and published by his descendant, Krehbiel, *Prairie Pioneer*. While skilled in wheat harvest, Krehbiel had no experience from Germany with chopping kafir corn. He enrolled himself in a club sponsored by his employer to master husking.

21. Henry King, "Picturesque Features of Kansas Farming," *Scribner's Monthly* 19, no. 1 (1879): 132.

22. Bernard Warkentin, Letter to David Goerz, July 29/11, 1872, Warkentin Papers, Bethel College.

23. My account of Warkentin's, Krehbiel's, and Schmidt's interactions is drawn from records preserved in the Bernhard Warkentin and Maurice L. Alden Papers, KSHS. Land records are preserved in the Atchison, Topeka & Santa Fe RR Co.—Historical Files, KSHS, boxes 567–68. A concise reminiscence of the overall terms of settlement is provided by the Santa Fe's chief agent for negotiating with the Mennonites, C. B Schmidt, "Reminiscences of Foreign Immigration Work for Kansas," *Transactions and Collections of the Kansas State Historical Society* 9 (1905): 485–97.

24. "Krehbiel Account," *Mennonite Life*, September 2003.

25. In addition to his autobiography, Krehbiel provides information about negotiating settlement in a travel narrative he prepared in 1872 for *Mennonitische Friedensbote*, a paper published in Milford Square, Pennsylvania, by the Eastern District Conference of the General Conference Mennonite Church.

26. C. B. Schmidt to Colonel A. S. Johnson, March 23, 1875. Old Correspondence and Papers. Land Department Sales to Mennonites; Immigration, KSHS, Box 308.

27. Mennonite history has been well narrated, albeit with an emphasis on loss of religious freedoms as the impetus for migration in the 1870s and with a tendency to ignore relations with surrounding communities of Russian peasants, Tatars, Doukhabors, and Jews. One early canonical account of Mennonite settlement is Smith, *The Story of the Mennonites, [by] C. Henry Smith.* Urry and Staples have noted and corrected this internalism, although disagreement persists over whether Russia's Great Reforms or dynamics internal to Mennonite communities were primary incentives to migration. On the impetus for migration, see James Urry, *None but Saints: The Transformation of Mennonite Life in Russia, 1789–1889* (Winnipeg: Hyperion Press, 1989); John R. Staples, *Cross-Cultural Encounters on the Ukrainian Steppe Settling the Molochna Basin, 1784–1861* (Toronto: University of Toronto Press, 2003). On the intercongregational rift, John B. Toews, *Perilous Journey: The Mennonite Brethren in Russia, 1860–1910* (Winnipeg: Kindred Press, 1988). On drought, see Moon, "In the Russians' Steppes."

28. Russian histories of agriculture have been dominated by a focus on serfdom, the principal mode of rural social and labor organization. Historians differ as to whether to regard the Mennonites as glorified serfs or settler farmers, but they indisputably enjoyed preferential terms of property ownership and resource control. On credit, see John R. Staples, "Johann Cornies, Money-Lending, and Modernization in the Molochna Mennonite Settlement, 1820s–1840s," *Journal of Mennonite Studies* 27 (2009): 109.

29. On the colonization of the steppes and the southern periphery, see Willard Sunderland, *Taming the Wild Field: Colonization and Empire on the Russian Steppe*, 1st ed. (Ithaca, NY: Cornell University Press, 2006); Nicholas Breyfogle, Abby Schrader, and Willard Sunderland, *Peopling the Russian Periphery: Borderland Colonization in Eurasian History*, 1st ed. (London: Routledge, 2007); David Moon, "Peasant Migration and the Settlement of Russia's Frontiers, 1550–1897," *Historical Journal* 40, no. 4 (1997): 859–93.

30. Sunderland, *Taming the Wild Field*, 86–87.

31. Roger P. Bartlett, *Human Capital: The Settlement of Foreigners in Russia, 1762–1804* (New York: Cambridge University Press, 1979); James Urry, *Mennonites, Politics, and Peoplehood Europe, Russia, Canada, 1525–1980* (Winnipeg: University of Manitoba Press, 2006).

32. On the career of Cornies, see Staples, *Cross-Cultural Encounters on the Ukrainian Steppe Settling the Molochna Basin*, 107–43; Urry, *None but Saints*, 119–30; Sunderland, *Taming the Wild Field*, 117–18.

33. On Cornies's reforms, see Urry, *None but Saints*; Moon, *Plough That Broke the Steppes*; on complaints about orchards, Daniel Schlatter, *Bruchstücke aus einigen Reisen nach dem südlichen Russland in den Jahren 1822 bis 1828: Mit besonderer Rücksicht auf die Nogaijen-Tartaren am Asowschen Meere* (Saint Gallen: Huber, 1836).

34. Staples, *Cross-Cultural Encounters on the Ukrainian Steppe Settling the Molochna Basin*, 50.

35. On Akkerman, see Staples, *Cross-Cultural Encounters on the Ukrainian Steppe Settling the Molochna Basin*, 112–14.

36. Moon, *Plough That Broke the Steppes*, 251.

37. On pastoralism as a flexible continuum of economic strategies, see Anatoly M. Khazanov, *Nomads and the Outside World*, trans. Julia Crookenden (New York: Cambridge University Press, 1984).

38. On relative yields and criticism of Nogai, see Staples, *Cross-Cultural Encounters*, 36–37. On Mennonite-Nogai economic relations, see Staples, *Cross-Cultural Encounters*, 142, and

Staples, "On Civilizing the Nogais: Mennonite-Nogai Economic Relations, 1825–1860," *Mennonite Quarterly Review*, April 2000, https://www.goshen.edu/mqr/pastissues/apr00staples.html, retrieved October 30, 2014.

39. This is the argument made by Staples, *Cross-Cultural Encounters on the Ukrainian Steppe Settling the Molochna Basin*, 45–86.

40. Moon, *Plough That Broke the Steppes*, 251–56.

41. Carol Belkin Stevens, *Soldiers on the Steppe: Army Reform and Social Change in Early Modern Russia* (DeKalb: Northern Illinois University Press, 1995); Arcadius Kahan and Richard Hellie, *The Plow, the Hammer, and the Knout: An Economic History of Eighteenth-Century Russia* (Chicago: University of Chicago Press, 1985), 50–55; Kelly Ann O'Neill, "Between Subversion and Submission: The Integration of the Crimean Khanate into the Russian Empire, 1783-1853" (PhD diss., Harvard University, 2006) 229.

42. Catherine seems to have had little plan to annex the peninsula initially, preferring instead to install the European educated and Russian friendly Tatar Sahin Giray as a puppet. After being rebuffed on the peninsula, he led a failed attempt at a coalition of Nogai Tatars to the north, temporarily pitting the Nogai against the Crimean Tatars. Thereafter, annexation became Russian policy. Alan W. Fisher, *Between Russians, Ottomans, and Turks: Crimea and Crimean Tatars* (Istanbul: Isis Press, 1998), 93–121.

43. Alessandro Stanziani, *After Oriental Despotism: Eurasian Growth in a Global Perspective* (London: Bloomsbury, 2014), 96.

44. Victor Ostapchuk and Svitlana Bilyayeva, "The Ottoman Black Sea Frontier at Akkerman Fortress," in *The Frontiers of the Ottoman World*, ed. A. C. S. Peacock (Oxford: Oxford University Press, 2009), 137–70, at p. 140.

45. Andrew Robarts, "A Plague on Both Houses? Population Movements and the Spread of Disease across the Ottoman-Russian Black Sea Frontier, 1768–1830s" (PhD diss., Georgetown University, 2010), 6–7. On the continuity of the Black Sea region as a unit of historical and political analysis, see Robarts, "Imperial Confrontation or Regional Cooperation? Bulgarian Migration and Ottoman-Russian Relations in the Black Sea Region, 1768–1830s," *Turkish Historical Review* 3, no. 2 (2012): 149–67.

46. Robarts, "Plague on Both Houses," 296.

47. On the carry trade and the expansion of Russian trade in the Black Sea, Patricia Ann McGahey Herlihy, "Russian Grain and Mediterranean Markets, 1774–1861" (PhD diss., University of Pennsylvania, 1978), and Vasilēs A Kardasēs, *Diaspora Merchants in the Black Sea: The Greeks in Southern Russia, 1775–1861* (Lanham, MD: Lexington Books, 2001), 109 ff.; on the opening of the port at Berdiansk, Staples, *Cross-Cultural Encounters on the Ukrainian Steppe Settling the Molochna Basin*, 123. On the British grain trade and the repeal of the Corn Laws, Susan Fairlie, "The Corn Laws and British Wheat Production, 1829–1876," *Economic History Review* 22, no. 1 (1969): 88–116.

48. On cultivated varieties and their markets, see Herlihy, "Russian Grain and Mediterranean Markets," 54–56, and Kardasēs, *Diaspora Merchants in the Black Sea*, 135. The traveler's account of Arnautka is Schlatter, *Bruchstücke aus einigen Reisen nach dem südlichen Russland in den Jahren 1822 bis 1828*. Touring Taurida in 1837, the Russian diplomat and industrialist Anatoly Demidov surveyed the local flora, including wild and cultivated wheats of the Crimean Peninsula. He noted *Aegilops cylindrica* and *triuncialis*, the former south of fortifications in Sevastopol, as well as *Triticum hybernum* in Crimea, which he reported cultivated mainly by German settlers from Neusatz to Kronenthal. Anatoliĭ Nikolaevich Demidov and Denis Auguste Marie Raffet, *Travels in Southern Russia and the Cri-*

mea; through Hungary, Wallachia, and Moldavia, during the Year 1837 (London: J. Mitchell, 1853), 161–62.

49. I thank John Staples for passing along this citation to the Harvest Report of 1812, Rossiiskii Gosudarstvennyi Isoricheskii arkhiv (RGIA) (Saint Petersburg), fond 1281, opis 11, delo 132, pp. 71–80.

50. On the demand for Sandomirka, see Herlihy, "Russian Grain and Mediterranean Markets," 56. On Mennonite experiments with new varietals, see Cornies correspondence with regional official and statistician Peter Köppen, *The Peter J. Braun Russian Mennonite Archive, 1803–1920: Documents Pertaining to the Molochna Mennonite Settlement in Southern Ukraine, Originally Assembled by Peter Braun during the Period 1917–1920; Re-Discovered in 1990 in the State Archives of the Odessa Region, Ukraine* (Odessa: State Archives of the Odessa Region, 1990–91).

51. Notably, Americans and Russians graded wheat differently, with Russians designating only durum wheat as hard, while Americans classified bread wheat as hard and soft. Rossiiskii Gosudarstvennyi Isoricheskii arkhiv (RGIA) (Saint Petersburg), 383, op. 29, 1837–38, d.609, II.37 ob.-8, in Moon, *Plough That Broke the Steppes*, 256, 288–89; Moon, "In the Russians' Steppes," 216–18.

52. Herlihy, "Russian Grain and Mediterranean Markets," 132–45. On requests to the US consul to procure American wheat, Skal'kovskii, *Population commerciale*, 5, in Herlihy, "Russian Grain and Mediterranean Markets," 135.

53. Consular Report, February 1, 1861, in Herlihy, "Russian Grain and Mediterranean Markets," 67.

54. Cornies's correspondence with Steven and regional official and statistician Peter Köppen is preserved in *The Peter J. Braun Russian Mennonite Archive, 1803–1920: Documents Pertaining to the Molochna Mennonite Settlement in Southern Ukraine, Originally Assembled by Peter Braun during the Period 1917–1920; Re-Discovered in 1990 in the State Archives of the Odessa Region, Ukraine*.

55. United States vice consul, Stephen Balli to Cass, Consular Report, November 29, 1860, quoted in Herlihy, "Russian Grain and Mediterranean Markets," 67.

56. On the Crimean khanate, Alan W. Fisher, *Between Russians, Ottomans, and Turks: Crimea and Crimean Tatars* (Istanbul: Isis Press, 1998); Brian Glyn Williams, *The Crimean Tatars: The Diaspora Experience and the Forging of a Nation* (Leiden: Brill, 2001), 39–72; Paul R. Magocsi, *This Blessed Land: Crimea and the Crimean Tatars* (Toronto: University of Toronto Press for the Chair of Ukrainian Studies, University of Toronto, 2014).

57. Alan W. Fisher, *The Crimean Tatars* (Stanford, CA: Hoover Institution Press, 1978), 23–24. Nobles consisted of bey, mirza, khan, and kapikulu governors.

58. Huri İslamoşlu-İnan, *State and Peasant in the Ottoman Empire: Agrarian Power Relations and Regional Economic Development in Ottoman Anatolia during the Sixteenth Century* (Leiden: Brill, 1994), 1–21.

59. Kelly Ann O'Neill, "Between Subversion and Submission," 265–315.

60. Azade-Ayşe Rorlich, *The Volga Tatars: A Profile in National Resilience* (Stanford, CA: Hoover Institution Press; Stanford University, 1986), 12–13 and 29–30, and Azade-Ayşe Rorlich, "Acculturation in Tatarstan: The Case of the Sabantui Festival," *Slavic Review* 2 (1982): 316–22.

61. Stanziani, *After Oriental Despotism*, 62–90.

62. Ostapchuk and Bilyayeva, "Ottoman Black Sea Frontier at Akkerman Fortress."

63. Herlihy, "Russian Grain and Mediterranean Markets," 26–30; Maria Guthrie and Matthew Guthrie, *A Tour, Performed in the Years 1795–6, through the Taurida, or Crimea: The*

Antient Kingdom of Bosphorus, the Once-Powerful Republic of Tauric Cherson, and All the Other Countries on the North Shore of the Euxine, Ceded to Russia by the Peace of Kainardgi and Jassy (London: Printed by Nichols and Son . . . for T. Cadell, Jun. and W. Davies . . . , 1802), 139–61.

64. Johann Wiebe, Letter from Bernard Warkentin, Newton, Kansas, requesting seed-grain and help in search for wheat varieties, 1900, File 3376, *Peter J. Braun Russian Mennonite Archive, 1803–1920*.

65. Warkentin to Carleton, November 9, 1900, RG 54, NARA, excerpted in Schellenberg summary, Warkentin Papers, Bethel College.

66. Letters of May 31, July 2, and November 9, 1900, RG 54, NARA, excerpted in Schellenberg summary, Warkentin Papers, Bethel College.

67. Moon, *Plough That Broke the Steppes*, 43.

CHAPTER FIVE

1. C. Henry Smith, *The Coming of the Russian Mennonites: An Episode in the Settling of the Last Frontier, 1874–1884* (Berne, IN: Mennonite Book Concern, 1927), 134.

2. Wesley Berg, "Bearing Arms for the Tsar: The Songs of the Germans in Russia," *Journal of Mennonite Studies* 17, no. 1 (1999): 180: the writer G. I. Kolesnikov. As early as 1848, an anonymous writer referred to Mennonite lands as an "oasis on the steppe," although he regarded it as a natural feature of the landscape rather than a result of human effort: Staples, *Cross-Cultural Encounters on the Ukrainian Steppe Settling the Molochna Basin*, 3.

3. "Kansas Farmers and Illinois Dairymen," *Atlantic Monthly* 44, no. 266 (1879): 717–25.

4. King, "Picturesque Features of Kansas Farming."

5. Krehbiel, *Prairie Pioneer*.

6. I. Dementyev, *A Trip to American Villages of Russian Mennonites in the State of Kansas* (from *A Tour of America in 1878*), *"Ustoi," Literary-Political Monthly*, no. 11 (Saint Petersburg: V. S. Balashev's Printing House, November 1, 1890): 114–37. I thank Sergey Shuvalov for his assistance with translation. Experiments with silk culture were abandoned by 1890, proving more labor intensive than wheat cultivation: Smith, *Coming of the Russian Mennonites*, 207.

7. E. V. Smalley, "The Isolation of Life on Prairie Farms," *Atlantic Monthly* 72, no. 431 (1893): 378–82.

8. Henry Van Dyke Jr., "The Red River of the North," *Harper's New Monthly Magazine* 60 (May 1880): 801–17.

9. *Topeka Commonwealth*, September 10, 1874, in Smith, *Coming of the Russian Mennonites*, 138.

10. Dementyev, "Trip to American Villages of Russian Mennonites."

11. Smith, *Coming of the Russian Mennonites*, 137.

12. An account of the crisis and succeeding controversy is provided in Warkentin's letters to Goerz, and in Krehbiel's autobiography.

13. Dementyev, "Trip to American Villages of Russian Mennonites."

14. Smith, *Coming of the Russian Mennonites*, 136; Dementyev, "Trip to American Villages of Russian Mennonites."

15. Dementyev, "Trip to American Villages of Russian Mennonites."

16. Smith, *Coming of the Russian Mennonites*, 137.

17. Ibid., 178–204.

18. John Warkentin, "The Mennonite Settlement of Southern Manitoba" (PhD diss., University of Toronto, 1960); Smith, *Coming of the Russian Mennonites*, 200.

19. Warkentin, "Mennonite Settlement of Southern Manitoba."

20. Ibid., 145.

21. Dementyev, "Trip to American Villages of Russian Mennonites."

22. On Hutterite communalism, see John W. Bennett, "Social Aspects of Sustainability and Common Property: Lessons from the History of the Hutterian Brethren," in *Human Ecology as Human Behavior: Essays in Environmental and Development Anthropology*, John W. Bennett (New Brunswick, NJ: Transaction Publishers, 1993); John W. Bennett, *Hutterian Brethren: The Agricultural Economy and Social Organization of a Communal People* (Stanford, CA: Stanford University Press, 1967); John A. Hostetler, *Hutterite Society* (Baltimore: Johns Hopkins University Press, 1974). On Hutterite migration to the Dakotas, Smith, *Coming of the Russian Mennonites*, 163–68.

23. On Amish social organization as a model for less input-intensive agriculture, see Deborah Stinner, Maurizio G. Paoletti, and B. R. Stinner, "In Search of Traditional Farm Wisdom for a More Sustainable Agriculture," *Agriculture Ecosystems and Environment* 27, nos. 1–4 (1989): 77–90. On the persistence of diversified farming amid the USDA's promotion of businessmen farmers, see Steven Dale Reschly and Katherine Jellison, *Production Patterns, Consumption Strategies, and Gender Relations in Amish and Non-Amish Farm Households in Lancaster County, Pennsylvania, 1935–1936* ([Berkeley, CA]: Agricultural History Society, 1993).

24. On Mennonite cooperative institutions, Smith, *Coming of the Russian Mennonites*, 203. On grain elevator operators, Cronon, *Nature's Metropolis*, 97–147; Charles Postel, *The Populist Vision* (New York: Oxford University Press), 116 ff.

25. Smith, *Coming of the Russian Mennonites*, 156; Cornelius Krahn, "Bernhard Warkentin," *Mennonite Year Book and Almanac* (1909): 24–26.

26. Burlington and Missouri Railroad Company pamphlet dated 1878, in Smith, *Coming of the Russian Mennonites*, 204.

27. Dementyev, "Trip to American Villages of Russian Mennonites."

28. Smith, *Coming of the Russian Mennonites*, 204.

29. Clark, Ball, and Martin, "Classification of American Wheat Varieties."

30. Dunbar, "Turkey Wheat."

31. Klippart, *Wheat Plant*, 496–98.

32. Moon, "In the Russians' Steppes"; Moon, *Plough That Broke the Steppes*, 203 ff.

33. King, "Picturesque Features of Kansas Farming."

34. Van Dyke, "Red River of the North," 801–18.

FIELD NOTES: "INDIGENOUS KNOWLEDGE"

1. Bonneuil and Fenzi, "From 'Genetic Resources' to 'Ecosystems Services.'"

2. Nikolai Vavilov, "The Problem of the Origin of the World's Agriculture in the Light of the Latest Investigations," Science at the Crossroads: Papers Presented to the International Congress of the History of Science and Technology Held in London from June 29th to July 3rd, 1931 by the delegates of the U.S.S.R., Frank Cass and Co., 1931; N. I. Vavilov and V. F. Dorofeev, *Origin and Geography of Cultivated Plants* (Cambridge: Cambridge University Press, 1992).

3. Paul Gepts, *Biodiversity in Agriculture: Domestication, Evolution, and Sustainability* (Cambridge: Cambridge University Press, 2012), 410–16.

4. On "genetic modernization" and the connections of animal and plant genetics to eugenics, see Bonneuil and Fenzi, "From 'Genetic Resources' to 'Ecosystems Services,'" and Michael Flitner, "Genetic Geographies: A Historical Comparison of Agrarian Modernization and Eugenic Thought in Germany, the Soviet Union, and the United States," *Geoforum* 34, no. 2 (2003): 175–85.

5. "Glottolog 2.3—Southeastern Iranian," Leipzig: Max Planck Institute for Evolutionary Anthropology, http://glottolog.org/resource/languoid/id/sout3156, accessed August 25, 2014.

6. Justin D. Faris, "Wheat Domestication: Key to Agricultural Revolutions Past and Future," in *Genomics of Plant Genetic Resources*, vol. 1, *Managing, Sequencing, and Mining Genetic Resources*, ed. Roberto Tuberosa, Andreas Graner, and Emile Frison (New York: Springer, 2014), 440–64.

7. Abena Dove Osseo-Asare, *Bitter Roots: The Search for Healing Plants in Africa* (Chicago: University of Chicago Press, 2014), 41–70.

8. Aichi Nagoya Protocol on Access and Benefit-Sharing (ABS), adopted October 29, 2010, at the tenth meeting of the Conference of the Parties (COP 10) to the Convention on Biological Diversity (CBD).

9. On RAFI, P. R. Mooney, *Seeds of the Earth: A Private or Public Resource?* (San Francisco; Ottawa: Institute for Food and Development Policy; Published by Inter Pares for the Canadian Council for International Co-operation and the International Coalition for Development Action [London], 1983); Cori Hayden, "From Market to Market: Bioprospecting's Idioms of Inclusion," *American Ethnologist* 30, no. 3 (2003): 359–71, doi:10.1525/ae.2003.30.3.359; C. Hayden, "Taking as Giving: Bioscience, Exchange, and the Politics of Benefit-Sharing," *Social Studies of Science* 37, no. 5 (2007): 729–58.

10. Rural Advancement Foundation International (RAFI), "Biopiracy Project in Chiapas, Mexico Denounced by Mayan Indigenous Groups," News Release—December 1, 1999, www.rafi.org; Brent Berlin and Elois Ann Berlin, "NGOs and the Process of Prior Informed Consent in Bioprospecting Research: The Maya ICBG Project in Chiapas, Mexico," *International Social Science Journal* 55, no. 178 (2003): 629–38; Brent Berlin and Elois Ann Berlin, "Community Autonomy and the Maya ICBG Project in Chiapas, Mexico: How a Bioprospecting Project That Should Have Succeeded Failed," *Human Organization* 63, no. 4 (2004): 472–86; Cori Hayden, *When Nature Goes Public: The Making and Unmaking of Bioprospecting in Mexico* (Princeton, NJ: Princeton University Press, 2003).

11. Vincristine is an Eli Lilly patented cancer drug derived from toxic alkaloids in the Madagascar periwinkle. Osseo-Asare, *Bitter Roots*, 41–70; Janice Harper, "The Not So Rosy Periwinkle: Political Dimensions of Medicinal Plant Research," *Ethnobotany Research and Applications* 3 (2005): 295–308.

12. On Taxol, Jordan Goodman and Vivien Walsh, *The Story of Taxol: Nature and Politics in the Pursuit of an Anti-Cancer Drug* (New York: Cambridge University Press, 2001). Paclitaxel (Taxol) is a cancer drug produced by Bristol-Myers Squibb derived from the bark of the Pacific yew tree (*Taxus brevifolia*), first identified by the National Cancer Institute in the 1960s.

13. Hämäläinen, "The Politics of Grass," 180.

CHAPTER SIX

1. Raw drug purchase inquiries and responses, Lloyd Brothers Pharmacists, Inc., 1870–1938, Collection 6, Series IV: Correspondence, Box 3, Lloyd Museum and Library, Cincinnati, Ohio.

2. On Russian thistle, Clinton L. Evans, *The War on Weeds in the Prairie West: An Environmental History* (Calgary: University of Calgary Press, 2002), 96–97, http://www.deslibris.ca/ID/402744; on *Elymus repens* and advocates of pasture grass research, Pauly, *Fruits and Plains*, 115–23; Frieda Knobloch, *The Culture of Wilderness Agriculture as Colonization in the American West* (Chapel Hill: University of North Carolina Press, 1996), 113–45.

3. "*Echinacea angustifolia* DC. blacksamson Echinacea," USDA Natural Resources Con-

servation Center, http://plants.usda.gov/java/profile?symbol=ECAN2, retrieved March 15, 2016; Michael A. Flannery, "From Rudbeckia to Echinacea: The Emergence of the Purple Cone Flower in Modern Therapeutics," *Pharmacy in History* 41, no. 2 (1999): 52–59.

4. Asahel Clapp uses the common name "black Samson" in his survey of indigenous medicinal plants prepared for the transactions of the American Medical Association: Asahel Clapp, "A Synopsis; or, Systematic Catalogue of the Indigenous and Naturalized, Flowering and Filicoid . . . Medicinal Plants of the United States: With Their Localities, Botanical and Medical References . . . Being a Report of the Committee on Indigenous Medical Botany and Materia Medica for 1850–51," *Transactions of the American Medical Association* 5 (1852): 698–906; see Flannery, "From Rudbeckia to Echinacea."

5. Judges 13–16.

6. On nomenclature: Kindscher, *Medicinal Wild Plants of the Prairie*, 84–85; Alma R. Hutchens, *Indian Herbalogy of North America* (Boston: Shambhala, distributed in the United States by Random House, 1991), 113; Daniel E. Moerman, *Native American Ethnobotany* (Portland, OR: Timber Press, 1998), 205–6.

7. Melvin R. Gilmore, *A Study in Ethnobotany of the Omaha Indians*, Nebraska State Historical Society, 1913; Daniel Moerman and University of Michigan–Dearborn, "Native American Ethnobotany: A Database of Foods, Drugs, Dyes, and Fibers of Native American Peoples, Derived from Plants," http://herb.umd.umich.edu/, accessed December 31, 2014. In his 1859 report on western botany for the secretary of war, V. Ferdinand Hayden noted that the abundant plant was "very pungent and used very effectively by the traders and Indians for the cure of the bite of the rattle snake." V. Ferdinand Hayden, "Botany," in *Preliminary Report of Explorations in Nebraska and Dakota in the Years 1855–'57*, by G. K. Warren (Washington, DC, 1875); Kindscher, *Medicinal Wild Plants of the Prairie*, 84–94.

8. Primary botanical texts listing *Rudbeckia purpurea* and *Echinacea* are William P. C. Barton et al., *A Flora of North America: Illustrated by Coloured Figures, Drawn from Nature* (Philadelphia. M. Carey & Sons, Chestnut Street, 1821); C. S. Rafinesque, *Medical Flora; or, Manual of the Medical Botany of the United States of North America: Containing a Selection of above 100 Figures and Descriptions of Medical Plants, with Their Names, Qualities, Properties, History, &c.: And Notes or Remarks on Nearly 500 Equivalent Substitutes: In Two Volumes: Volume the First, A-H with 52 Plates [Volume the Second, with 48 Plates]*, 1828; Asa Gray and William Starling Sullivant, *A Manual of the Botany of the Northern United States from New England to Wisconsin and South to Ohio and Pennsylvania Inclusive: (the Mosses and Liverworts by Wm. S. Sullivant) Arranged according to the Natural System* (Boston: J. Munroe, 1848). Barton used Linneaus's identification of *Rudbeckia*, whereas Rafinesque called it *Helichroa*. Gray adopted Mönch's nomenclature. On the varied references to *purpurea, pallida, and angustifolia* in Euro-American medical botany, see Flannery, "From Rudbeckia to Echinacea."

9. A synthetic history on the professionalization of American medicine is Paul Starr, *The Social Transformation of American Medicine* (New York: Basic Books, 1982). For revisions and reconsiderations, Timothy S. Jost, Mark Schlesinger, and Keith Wailoo, "Transforming American Medicine: A Twenty-Year Retrospective on the Social Transformation of American Medicine," *Journal of Health Politics, Policy and Law* 29, no. 4/5 (2004).

10. John S. Haller, *A Profile in Alternative Medicine: The Eclectic Medical College of Cincinnati, 1845–1942* (Kent, OH: Kent State University Press, 1999), 1–26.

11. Michael A. Flannery, *John Uri Lloyd: The Great American Eclectic* (Carbondale: Southern Illinois University Press, 1998); D. B. Worthen, "John Uri Lloyd, 1849–1936: Wizard of American Plant Pharmacy," *Journal of the American Pharmacists Association* 49, no. 2 (2009). Diary

and expense books, 1864–68, Box 1, John Uri Lloyd Papers, 1849–1936, Collection No. 1, Lloyd Library and Museum, Cincinnati, Ohio.

12. Worthen, "John Uri Lloyd"; Flannery, *John Uri Lloyd*.

13. Krehbiel, *Prairie Pioneer*. On Hahnemann, see John S. Haller, *The History of American Homeopathy: The Academic Years, 1820–1935* (New York: Pharmaceutical Products Press, 2005). On German pharmacological tradition in North America, see Renate Wilson, *Pious Traders in Medicine: A German Pharmaceutical Network in Eighteenth-Century North America* (University Park: Pennsylvania State University Press, 2000).

14. The London, Edinburgh, and US Pharmacopoeia listed botanical remedies alongside mineral medicines and chemical preparations, canonizing centuries of popular medical practice and adding a growing list of New World plants to the list. Edward Kremers, George Urdang, and Glenn Sonnedecker, *Kremers and Urdang's History of Pharmacy* (Madison, WI: American Institute for the History of Pharmacy [AIHP], 1976).

15. On the transit of plant drugs, see for example, Crosby, *Ecological Imperialism*; Londa Schiebinger, *Plants and Empire: Colonial Bioprospecting in the Atlantic World* (Cambridge: Cambridge University Press, 2004); Cook, *Matters of Exchange*. On potential plant drugs from the North American colonies, see David Cowen, "The British North American Colonies as a Source of Drugs," Die Vortrage der Hauptversammlung, Bd 28 (Stuttgart: Wissenschaftliche Verlagsgesellschaft, 1966).

16. Glenn Sonnedecker, *The Founding of the Pharmacopeia: A Series of Three Articles* (Rockville, MD: Reprinted by the United States Pharmacopeia with the permission of the American Institute of the History of Pharmacy, 1995).

17. Benjamin Rush, "An Inquiry into the Natural History of Medicine among the Indians of North America: And a Comparative View of Their Diseases and Remedies with Those of Civilized Nations," *Medical Inquiries and Observations* 1 (1794): 9–77. For a more sensitive account of Native American medicine from the same period, note, for example, the first two titles the Lloyd Brothers chose to print in their series of Bulletins on the history of plant drugs: Reproduction Series No. 1, "Collections for an Essay Towards a Materia Medica of the United States" (1798, 1804), Benjamin Smith Barton, 1900. Bulletin No. 2 Reproduction Series No. 2, "The Indian Doctor's Dispensatory Being Father Smith's Advice Respecting Diseases and Their Cure" (1812), Peter Smith, 1901. A general account of Anglo-American assessments of Native American medicine in the early Republic is Virgil J. Vogel, *American Indian Medicine* (Norman: University of Oklahoma Press, 1970), 36–110.

18. Terry M. Parssinen, *Secret Passions, Secret Remedies: Narcotic Drugs in British Society, 1820–1930* (Philadelphia: Institute for the Study of Human Issues, 1983), 15–20. For a typical inventory held by an eighteenth-century apothecary representing imported botanical medicines, see David L. Cowen, "A Store Mixt, Various, Universal," *Journal of the Rutgers University Library* 25 (December 1961): 1–9.

19. Edwin Troxell Freedly, *A Treatise on the Principal Trades and Manufacturers of the United States*, 1856; Silliman's journal; Glenn Sonnedecker, "Contributions of the Pharmacy Profession toward Controlling the Quality of Drugs in the Nineteenth Century," in *Safeguarding the Public: Historical Aspects of Medicinal Drug Control*, ed. John B. Blake (Baltimore: Johns Hopkins University Press, 1970), 98–99.

20. Saul Jarcho, "Drugs Used at Hudson Bay in 1730," *Bulletin of the New York Academy of Medicine* 47, no. 7 (1971): 838–42.

21. On pharmacy and therapeutics on slave plantations, see Sharla M. Fett, *Working Cures: Healing, Health, and Power on Southern Slave Plantations* (Chapel Hill: University of North

Carolina Press, 2002), and Todd L. Savitt, *Medicine and Slavery: The Diseases and Health Care of Blacks in Antebellum Virginia* (Urbana: University of Illinois Press, 1978).

22. A useful study is W. Simpson, "The Names of Medicinal Plants of Commercial Value That Are Gathered in North Carolina: Their Value, and Relative Amount Sold in This Country and Exported," *Proceedings of the American Pharmaceutical Association at the Annual Meeting* 42 (1894): 210–20. A study of the history and persistence of Appalachian medicinal folkways using later twentieth-century ethnographic data is John K. Crellin's preliminary study, "Traditional Medicine in Southern Appalachia and Some Thoughts for the History of Medicinal Plants," in *Botanical Drugs of the Americas in the Old and New Worlds: Invitational Symposium at the Washington-Congress*, Stuttgart, 1983, and the final study based on a decade of fieldwork: John K. Crellin and Jane Philpott, *Trying to Give Ease: Tommie Bass and the Story of Herbal Medicine* as well as the companion *Reference Guide to Medicinal Plants: Herbal Medicine Past and Present*, both (Durham, NC: Duke University Press, 1997). On domestic botanic medicine, see Judith Sumner, *American Household Botany: A History of Useful Plants, 1620–1900* (Portland, OR: Timber Press, 2004), 62, 241.

23. John Tennent, *Every Man His Own Doctor; or, the Poor Planter's Physician, the First Such American Guide to Domestic Practice*, 2nd ed. (Williamsburg, VA; Annapolis, MD: William Parks, 1734); William Buchan, *Domestic Medicine* (Edinburgh, 1769). In English practice, the bible of these was Nicholas Culpeper's *Complete Herbal* and *English Physician*, first published separately in 1652 and 1652, respectively, and thereafter in numerous editions: Nicholas Culpeper, *Culpeper's Complete Herbal: Consisting of a Comprehensive Description of Nearly All Herbs with Their Medicinal Properties and Directions for Compounding the Medicines Extracted from Them* (London: W. Foulsham, 1950).

24. John Gunn, *Domestic Medicine: A Facsimile of the First Edition [1830] with an Introduction by Charles E. Rosenberg* (Knoxville: University of Tennessee Press, 1986).

25. "Patent" is a misnomer, since in fact the remedies were not patented (requiring disclosure in exchange for a limited monopoly on production) but kept secret—or at least, a full list of ingredients was rarely published on the label.

26. On sarsaparilla and Swaim's Panacea, see, for example: William Bromfield, *An Account of the English Nightshades, And Their Effects . . . Also Practical Observations on the Use of Corrosive Sublimate, and Sarsaparilla; on the different Effects of Mercury Crude, and When Prepared by Chemistry . . .* (London: Printed for R. Baldwin, in Paternoster-Row, and G. Woodfall, Charing-Cross, 1757); John Gerard, *The Herball, or Generall Historie of Plantes*, 1663, revised and enlarged by Thomas Johnson, based on the original 1597 edition (Mineola, New York: Dover Publications, 1976); Edmund Morgan, "John White and the Sarsaparilla," *William and Mary Quarterly* (July 1957): 414–17; John Pope, *On the Comparative Virtues of Different Kinds of Sarsaparilla, by Mr. John Pope, Communicated by Mr. Earle, From the 12th vo of the Medico-chirurgical transactions, published by the medical and chirurgical society of London, to which is added an appendix (not included in the original comm.) on Cinchonine, Quinine, and other New Vegetable Principles* (London: E. Bridgewater, 1823); Reports of the Medical Society of the City of New York on Nostrums, or Secret Medicines. Part 1. Published by the Order of the Society, Under the Direction of the Committee on Quack Medicines (New York, 1829); *Smilax, Linnaeus, pseud., Sarsaparilla, and sarsaparilla so called: a popular analysis of a popular medicine, its nature, properties, and uses, how to insure its success as a remedy, the most approved forms, and the various phases of disease in which it may be advantageously employed* (London: Aylott, 1854).

27. On the marketing techniques of patent medicine makers: James Harvey Young, *The*

Toadstool Millionaires: A Social History of Patent Medicines in America before Federal Regulation (Princeton, NJ: Princeton University Press, 1961); A. Walter Bingham, *The Snake Oil Syndrome: Patent Medicine Advertising* (Hanover, MA: Christopher Publishing House, 1994); Martin Kaufman, "'Step Right Up Ladies and Gentlemen': Patent Medicines in Nineteenth Century America," *American History Illustrated*, August 1981.

28. Judith Sumner, *American Household Botany* (Portland, OR: Timber Press, 2012), 257–58; Susan Strasser, "Commodifying Lydia Pinkham: A Woman, a Medicine, and a Company in a Developing Consumer Culture," Working Paper 32, ESRC/AHRB Cultures of Consumption Programme, http://www.consume.bbk.ac.uk/publications.html.

29. George Winston Smith, *Medicines for the Union Army: The United States Army Laboratories during the Civil War* (Madison, WI: American Institute of the History of Pharmacy, 1962); Michael A. Flannery, *Civil War Pharmacy: A History of Drugs, Drug Supply and Provision, and Therapeutics for the Union and Confederacy* (New York: Pharmaceutical Products Press, 2004).

30. Alex Berman and Michael Flannery, *America's Botanico-Medico Movements: Vox Populi* (New York: Pharmaceutical Products Press, 2001); John S. Haller, *American Medicine in Transition, 1840–1910* (Urbana: University of Illinois Press, 1981); Haller, *History of American Homeopathy*.

31. Samuel Thomson, *A Narrative of the Life and Medical Discoveries of Samuel Thomson; Containing an Account of His System of Practice, and the Manner of Curing Disease with Vegetable Medicine, upon a Plan Entirely New; to Which Is Added an Introduction to His New Guide to Health, or Botanic Family Physician, Containing the Principles upon Which the System Is Founded, with Remarks on Fevers, Steaming, Poison, &c.* (New York: Arno Press, 1972); John S. Haller, *The People's Doctors: Samuel Thomson and the American Botanical Movement, 1790–1860* (Carbondale: Southern Illinois University Press, 2000).

32. Flannery, *Civil War Pharmacy*; George Winston Smith and American Institute of the History of Pharmacy, *Medicines for the Union Army: The United States Army Laboratories during the Civil War* (Madison, WI: American Institute of the History of Pharmacy, 1962).

33. For example, Starr, *Social Transformation of American Medicine*; Berman and Flannery, *America's Botanico-Medical Movements: Vox Populi*.

34. On the contest for authority in the mid-nineteenth century, see John Harley Warner, "Power, Conflict, and Identity in Mid-Nineteenth-Century American Medicine: Therapeutic Change at the Commercial Hospital in Cincinnati," *Journal of American History* 73, no. 4 (1987): 934–56.

35. Charles Rosenberg, "The Therapeutic Revolution: Medicine, Meaning, and Social Change in Nineteenth-Century America," *Perspectives in Biology and Medicine* 20 (1977): 458–506; John Harley Warner, *The Therapeutic Perspective: Medical Practice, Knowledge, and Identity in America, 1820–1885* (Cambridge, MA: Harvard University Press, 1986).

36. Haller, *Profile in Alternative Medicine*; John S. Haller, *Medical Protestants: The Eclectics in American Medicine, 1825–1939* (Carbondale: Southern Illinois University Press, 1994).

37. Kindscher, *Medicinal Wild Plants of the Prairie*, 90; Sumner, *American Household Botany*, 256; Flannery, "From Rudbeckia to Echinacea," 53–54.

38. Libradol, Box 7, Collection No. 6, Lloyd Brothers Pharmacists, Inc., 1870–1938, Lloyd Library and Museum, Cincinnati, Ohio.

39. Worthen, "John Uri Lloyd"; Flannery, *John Uri Lloyd*.

40. Only two files on plant acquisition for Lloyd Brothers survive, both from the year 1903. These relate exclusively to supplies of *Lobelia* and *Echinacea*. As a result, most of what we know about the business's supply networks is a result of inference and extrapolation. How-

ever, these files are noteworthy because they indicate far-flung correspondence networks and sites of inquiry rather than simple orders from major drug importers. All subsequent discussion of Lloyd's raw drug procurement comes from these files unless otherwise noted. Raw drug purchase inquiries and responses, Lloyd Brothers Pharmacists, Inc., 1870–1938, Collection 6, Series IV: Correspondence, Box 3, Lloyd Museum and Library, Cincinnati, Ohio.

41. H. W. Fetter, "The Newer Materia Medica: Echinacea," *Eclectic Medical Journal* 58 (1898): 79–89; Kindscher, *Medicinal Wild Plants of the Prairie*, 90.

42. John K. Crellin and Jane Philpott, *Trying to Give Ease: Tommie Bass and the Story of Herbal Medicine*, reprint ed. (Durham, NC: Duke University Press Books, 1997).

43. L. E. Sayre, "Therapeutical Notes and Description of Parts of Medicinal Plants Growing in Kansas," *Transactions of the Annual Meetings of the Kansas Academy of Science* 16 (1897): 86; Kindscher, *Medicinal Wild Plants of the Prairie*, 90–93.

44. US Federal Census, Twelfth Census of the United States, Schedule No. 1—Population: Paola, Miami County, Kansas, 1900.

45. Fett, *Working Cures: Healing, Health, and Power on Southern Slave Plantations*, 104.

46. John Uri Lloyd, "A Treatise on Echinacea," *Lloyd Brothers Drug Treatises* 30 (1917); Lloyd, "Vegetable Drugs Employed by American Physicians," *Journal of American Pharmaceutical Association* 1 (November 1912): 1228–41, latter noted in Flannery, "From Rudbeckia to Echinacea," 55.

47. One of the most enduring accounts of the power of regulators in curbing nineteenth-century "quack medicines" is Young, *Toadstool Millionaires*. A synthetic account of the AMA's consolidation is Starr, *Social Transformation of American Medicine*.

48. Abraham Flexner and Henry S. Pritchett, *Medical Education in the United States and Canada: A Report to the Carnegie Foundation for the Advancement of Teaching* (New York City, 1910). Alex Berman identifies 1910 as a pivotal year in the history of pharmacy in light of these three major developments. Berman and Flannery, *America's Botanico-Medical Movements*, 150–51.

49. Council on Pharmacy and Chemistry, "Echinacea Considered Valueless," *American Medical Association Journal* 53 (1909): 1836.

50. L. E. Sayre, "Echinacea Roots," *Transactions of the Kansas Academy of Science* 19 (1903): 209–13; Dana M. Price and Kelly Kindscher, "One Hundred Years of *Echinacea angustifolia* Harvest in the Smoky Hills of Kansas, USA," *Economic Botany* 61, no. 1 (2007): 86–95; Sumner, *American Household Botany*. A recent resurgence of interest in *Echinacea*'s immune boosting properties have revived concerns about conservation.

51. "America's Great Treasure," Box 39, Collection No. 1, John Uri Lloyd Papers.

52. Alfred Chandler, *Shaping the Industrial Century: The Remarkable Story of the Modern Chemical and Pharmaceutical Industries* (Cambridge, MA: Harvard University Press, 2005); Walter Sneader, *Drug Discovery: A History* (Hoboken, NJ: Wiley, 2005); Joseph M. Gabriel, *Medical Monopoly: Intellectual Property Rights and the Origins of the Modern Pharmaceutical Industry* (Chicago: University of Chicago Press, 2014); Dominique A. Tobbell, *Pills, Power, and Policy: The Struggle for Drug Reform in Cold War America and Its Consequences* (Berkeley: University of California Press; New York: Millbank Memorial Fund, 2012).

53. Harvard University anthropologist Richard Evans Schultes is often regarded as the father of ethnobotany for his studies into hallucinogenic and spiritual plant use among indigenous peoples of the Americas from the 1930s. While Schultes's studies motivated generations of students, many of whom were active in the USDA Bureau of Plant Industry initiated by Mark Carleton, the exploitation of plants and knowledge about them has a much longer his-

tory. And phytochemistry, with its emphasis on isolation and structure determination, differs significantly from botany in its orientation toward the natural world and the field of human knowledge that has developed in relation to it.

54. Price and Kindscher, "One Hundred Years of *Echinacea angustifolia* Harvest in the Smoky Hills of Kansas."

55. "Lloyd Brothers, Pharmacists, Inc.," http://www.lloydlibrary.org/history/lloyd%20pharmacy.html, retrieved November 4, 2011. On German research on *Echinacea* since the 1920s, see Flannery, "From Rudbeckia to Echinacea," 56–58.

56. Worthen, "John Uri Lloyd."

57. John Uri Lloyd and J. A. Knapp, *Etidorhpa; or, the End of Earth: The Strange History of a Mysterious Being and the Account of a Remarkable Journey as Communicated in Manuscript to Llewellyn Drury Who Promised to Print the Same, but Finally Evaded the Responsibility Which Was Assumed* (Cincinnati: Robert Clarke, 1896).

CHAPTER SEVEN

1. Theodore Roosevelt, *Hunting Trips of a Ranchman: Sketches of Sport on the Northern Cattle Plains* (New York: G. P. Putnam's Sons, 1885).

2. Jacoby, *Crimes against Nature*; Sellars, *Preserving Nature in the National Parks.*

3. Donna Haraway, "Teddy Bear Patriarchy: Taxidermy in the Garden of Eden, New York City, 1908–1936," *Social Text* 11 (Winter 1984): 20–64.

4. John Uri Lloyd, "A Fragment of Unwritten History concerning the Spanish War Tax," Box 39, Collection No. 1, John Uri Lloyd Papers, Lloyd Library and Museum, Cincinnati, Ohio. All narrative of the war tax attributed to Lloyd hereafter is from this essay unless otherwise noted. Variations on this essay also appear in Box 18 of the same collection.

5. T. F. Carmody and F. M. Peasley, *United States War Revenue Law of June 13, 1898: With Annotated References to the Rulings on the Same by the Treasury Department from the Passage of the Act to December 29, 1898* (Waterbury, CT: Dissell, 1899).

6. John Uri Lloyd, "Wilson-Lloyd correspondence, 1898–1900," Box 18, Collection No. 1, John Uri Lloyd Papers. Lloyd labeled them Exhibits A and C, with multiple handwritten drafts attached.

7. A synthetic account of the AMA's consolidation is Starr, *Social Transformation of American Medicine*; Berman and Flannery, *America's Botanico-Medical Movements*, 150–51.

8. "Treasury Decisions under Internal Revenue Laws of the United States: Reformatted from the Original and Including, Compilation of Decisions Rendered by the Commissioner of Internal Revenue Under the War-Revenue Act of June 13, 1898." Washington, DC: Government Printing Office, 1899.

9. John Uri Lloyd, "The War Tax," Box 18, Collection No. 1, John Uri Lloyd Papers.

10. Worthen, "John Uri Lloyd," summarizes his accomplishments succinctly; Flannery, *John Uri Lloyd.*

11. John Uri Lloyd and J. Augustus Knapp, *Etidorhpa, or the End of the Earth. The strange history of a mysterious being and the account of a remarkable journey as communicated in manuscript to L. Drury, who promised to print the same, but finally evaded the responsibility, which was assumed by J. U. Lloyd. With many illustrations by J. A. Knapp* (Cincinnati: J. U. Lloyd, 1895).

12. Flannery, *John Uri Lloyd*, 117 ff.

13. Raw drug purchase inquiries and responses, Lloyd Brothers Pharmacists, Inc., 1870–1938, Collection 6, Series IV: Correspondence, Box 3, Lloyd Museum and Library, Cincinnati, Ohio.

14. On grave robbing, Michael Sappol, *A Traffic of Dead Bodies: Anatomy and Embodied Social Identity in Nineteenth-Century America* (Princeton, NJ: Princeton University Press, 2002). On the Eclectics' engagement in this practice, Harvey Wickes Felter, *History of the Eclectic Medical Institute* (1902) 16–17.

15. Plato and Benjamin Jowett, *Plato's the Republic* (New York: Modern Library, 1941), Book VII.

16. Dorothy Ross, *The Origins of American Social Science*, Ideas in Context (Cambridge: Cambridge University Press, 1991).

17. Roosevelt, *Hunting Trips of a Ranchman.*

18. *An elk-hunt at Two Ocean Pass.* [1893?]. Theodore Roosevelt Collection. HCL_1834_1110i. Houghton Library, Harvard University, http://www.theodorerooseveltcenter.org/en/Research/Digital-Library/Record.aspx?libID=0286523. Theodore Roosevelt Digital Library. Dickinson State University.

19. Haraway, "Teddy Bear Patriarchy." Susan Schrepfer has also emphasized the varied attitudes toward nature that informed Progressive conservationists: Susan R. Schrepfer, *Nature's Altars: Mountains, Gender, and American Environmentalism* (Lawrence: University Press of Kansas, 2005).

20. "Old Dog Turk," John Uri Lloyd Papers, 1849–1936, Collection 1, Series II: Autobiographical Writings, Box 2.

21. "The Big Bone Country of Kentucky," John Uri Lloyd Papers, 1849–1936, Collection 1, Series XVII: Literary Manuscripts, Sub-Series J: Articles, Box 39.

CHAPTER EIGHT

1. Inscription on verso of a grain of rice, presented to Henry Ford in 1930 by Nassib Makarem: "First three Model T cars Henry Ford produced."

2. Fernand Braudel, "Alimentation et catégories de l'histoire," *Annales Économies, Sociétés, Civilisations* 16, no. 4 (1961): 723–28; Andrew Shryock, Daniel Lord Smail, and Timothy K. Earle, *Deep History: The Architecture of Past and Present* (Berkeley: University of California Press, 2011).

3. Sheik Nassib Makarem's art was preserved by his son Sami Makarem, a professor of Arabic literature, Islamic thought, and Sufism at American University, Beirut, in Sami Makarem, *The Calligraphy of Sheik Nassib Makarem* (Beirut: American University, Beirut, [n.d.]), English introduction: http://almashriq.hiof.no/ddc/projects/arabic/calligraphy/english.html; "The Fine Writing," http://www.nassibmakarem.org/fineWriting.html. Samples of Nassib Makarem's calligraphy, although not on grains, are included in the Personal Series of the American Druze Heritage collection, Julia Makarem Papers, 1930–2006, Box 1, Michigan Historical Collections, Bentley Historical Library, University of Michigan, Ann Arbor. I am grateful to Julia's children, Sahar Makarem Kadi, and Samir Makarem, for helping me locate photographs of Nassib Makarem's calligraphy, taken some twenty years ago at the hospital of American University, Beirut, using a microscope camera.

4. Robert Hooke and R. T Gunther, *Micrographia; or, Some Physiological Descriptions of Minute Bodies Made by Magnifying Glasses: With Observations and Inquiries Thereupon* (1665; New York: Dover Publications, 1961), Observation XVIII. I'm grateful to Courtney Weiss Smith for discussing the *Micrographia* with me.

5. Klippart, *Wheat Plant*, 27.

6. "FAOStat," n.d., http://faostat.fao.org/site/339/default.aspx., retrieved April 23, 2014.

7. On grading systems and futures markets, see Cronon, *Nature's Metropolis*; Jonathan Ira

Levy, “Contemplating Delivery: Futures Trading and the Problem of Commodity Exchange in the United States, 1875–1905,” *American Historical Review* 111, no. 2 (2006): 307–35.

8. On commensurability, see Wendy Nelson Espeland and Mitchell L Stevens, “A Sociology of Quantification,” *European Journal of Sociology* 49, no. 3 (2008).

9. Paul J. Crutzen and Eugene F. Stoermer, “The ‘Anthropocene,’” *International Geosphere-Biosphere Programme Newsletter* 41 (May 2000): 17–18.

10. On periodization, Julia Adeney Thomas, “History and Biology in the Anthropocene: Problems of Scale, Problems of Value,” *American Historical Review* 119, no. 5 (2014): 1587–607; Fredrik Albritton Jonsson, “Anthropocene Blues: Abundance, Energy, Limits,” in “The Imagination of Limits: Exploring Scarcity and Abundance,” ed. Frederike Felcht and Katie Ritson, RCC Perspectives 2015, no. 2, 55–63; Dipesh Chakrabarty, “The Climate of History: Four Theses,” *Critical Inquiry* 35, no. 2 (2009): 197–222.

11. Chakrabarty, “Climate of History,” 220.

12. M. J. S. Rudwick, *Bursting the Limits of Time: The Reconstruction of Geohistory in the Age of Revolution* (Chicago: University of Chicago Press, 2005); William R. Farrand, “Origins of Quaternary-Pleistocene-Holocene Stratigraphic Terminology,” in *Establishment of a Geologic Framework for Paleoanthropology*, ed. Léo F. Laporte, Geological Society of America, and History of Geology Division (Boulder, CO: Geological Society of America, 1990).

13. Bruce G. Trigger, *A History of Archaeological Thought* (New York: Cambridge University Press, 1989).

14. Joy McCorriston, “Domestication and the Dialectic: Archaeobotany and the Future of the Neolithic Revolution in the Near East,” in *From Foragers to Farmers: Papers in Honour of Gordon C. Hillman*, ed. Andrew S. Fairbairn and Ehud Weiss (Oxford, UK; Oakville, CT: Oxbow Books, 2009). On Childe, Bruce G. Trigger, *Gordon Childe: Revolutions in Archaeology* (New York: Columbia University Press, 1980). On Morgan, Lewis Henry Morgan, *Ancient Society* (Cambridge, MA: Belknap Press of Harvard University Press, 1964).

15. E.g., Andrew S. Fairbairn and Ehud Weiss, eds., *From Foragers to Farmers: Papers in Honour of Gordon C. Hillman* (Oxford, UK; Oakville, CT: Oxbow Books, 2009).

16. Rachel Brenchley et al., “Analysis of the Bread Wheat Genome Using Whole-Genome Shotgun Sequencing,” *Nature* 491, no. 7426 (2012): 705–10.

17. Daniel Lord Smail, *On Deep History and the Brain* (Berkeley: University of California Press, 2008); Shryock, Smail, and Earle, *Deep History*.

18. Gilles Deleuze and Félix Guattari, *A Thousand Plateaus: Capitalism and Schizophrenia* (Minneapolis: University of Minnesota Press, 1987). But perhaps they didn’t give trees their due: Catrina Sandilands imagines, for example, an elephantine oak tree whose encasements of death enable the shoots of green that issue forth from its branches: Catriona Sandilands, “Botanically Queer,” Heyman Center for the Humanities, New York, 2014.

19. William Connolly draws on the insights of complexity theory in biology to suggest different “onto-theological” images of time. William E. Connolly, *A World of Becoming* (Durham, NC: Duke University Press, 2010).

20. Peter Galison, “Einstein’s Clocks: The Place of Time,” *Critical Inquiry* 26, no. 2 (2000): 355–89.

21. E. P. Thompson, “Time, Work-Discipline, and Industrial Capitalism,” *Past and Present* 38, no. 1 (1967): 56–97.

22. For example, Harry Braverman, *Labor and Monopoly Capital: The Degradation of Work in the Twentieth Century* (New York: Monthly Review Press, 1975).

23. Fernand Braudel, *The Mediterranean and the Mediterranean World in the Age of Philip II*

(New York: Harper and Row, 1972); Fernand Braudel, *Capitalism and Material Life, 1400–1800* (New York: Harper and Row, 1973).

24. For example, Steven Topik and Allen Wells, *Global Markets Transformed: 1870–1945* (Cambridge, MA: Belknap Press of Harvard University Press, 2014); Jennifer Bair, *Frontiers of Commodity Chain Research* (Stanford, CA: Stanford University Press, 2009). On world systems theory, e.g., Immanuel Wallerstein, *The Modern World-System* (New York: Academic Press, 1974).

25. E.g., Michel Foucault, *The Order of Things: An Archaeology of the Human Sciences* (New York: Pantheon Books, 1971); Walter Benjamin, "Theses on the Philosophy of History," in *Illuminations* (New York: Harcourt, Brace and World, 1968).

26. Reinhart Koselleck, *Futures Past: On the Semantics of Historical Time* (New York: Columbia University Press, 2004). Koselleck believed revolutionary will culminated in fascism, shifting from the French Revolution's failures to a rumination on the Third Reich, but we can use his analysis to think about the requirements of modernity rather than political outcomes. Koselleck, *Futures Past*, 22–25.

27. Brian W. Ogilvie, *The Science of Describing Natural History in Renaissance Europe* (Chicago: University of Chicago Press, 2006), 139–264.

28. On type specimens and the compression of diversity in classification, Lorraine Daston, "Type Specimens and Scientific Memory," *Critical Inquiry* 31, no. 1 (2004): 153–82, doi:10.1086/427306; Cooper, *Inventing the Indigenous.*

29. On acclimatization as the paradigmatic colonial science, see Michael A. Osborne, "Acclimatizing the World: A History of the Paradigmatic Colonial Science," *Osiris*, 2nd series, 15 (January 1, 2000): 135–51.

30. Ogilvie, *Science of Describing Natural History in Renaissance Europe*, 165.

31. E-mail from Josephine Piggin, ICARDA staff botanist, April 12, 2014.

32. Nigel Maxted et al., "Gap Analysis: A Tool for Complementary Genetic Conservation Assessment," *Diversity and Distributions* 14, no. 6 (2008): 1018–30.

33. Klippart, *Wheat Plant*, ix–xv.

34. Jeffrey Abt, *American Egyptologist: The Life of James Henry Breasted and the Creation of His Oriental Institute* (Chicago: University of Chicago Press, 2011), chap. 5.

35. Justin D. Faris, "Wheat Domestication: Key to Agricultural Revolutions Past and Future."

36. On Robinson's influences, see ibid. The perspective on Breasted's sympathies benefits from conversations with Martin Jones of the George Pitt Rivers Laboratory for Bioarchaeology at the University of Cambridge.

37. Klippart, *Wheat Plant*, ix–xv.

38. On the corporatization of breeding, see Kloppenburg, *First the Seed.*

39. For a succinct diagram of Borlaug and Orville Vogel's field trials with winter and spring wheat semidwarfs, see Mahabal Ram, *Plant Breeding Methods* (Prentice-Hall India Learning, 2014), 22–60.

40. ICARDA Press Release, "ICARDA safeguards world heritage of genetic resources during the conflict in Syria," http://www.icarda.org/update/press-release-icarda-safeguards-world-heritage-genetic-resources-during-conflict-syria#sthash.cDr1bw13.4dGsVUxG.dpuf, retrieved September 28, 2015. Widely reported in more sensationalist terms: e.g., "Syrian civil war causes the first ever withdrawal from the 'doomsday bank,'" http://www.independent.co.uk/news/world/middle-east/syrian-civil-war-causes-the-first-ever-withdrawal-from-the-doomsday-bank-10512015.html, retrieved September 28, 2015; "Syria's civil war prompts first

'Doomsday Vault' withdrawal," http://edition.cnn.com/2015/09/24/world/norway-seed-vault/index.html, retrieved September 28, 2015.

41. On the politics of public sector breeding, see, variously, Cullather, *Hungry World*; Jonathan Harwood, *Europe's Green Revolution and Others Since: The Rise and Fall of Peasant-Friendly Plant Breeding* (London: Routledge, 2012); Perkins, *Geopolitics and the Green Revolution*; Daniel Immerwahr, *Thinking Small: The United States and the Lure of Community Development* (Cambridge, MA: Harvard University Press, 2015).

42. Charles Montgomery, the late dean of decorative arts at Yale University, translated the practice of connoisseurship into a fourteen-point "homework of study" intended to cultivate subjective habits of mind. The analytics of material culture studies can help us understand seeds in a different way. Associated with the discriminating gaze of Renaissance gentlemen perhaps also partial to cabinets of naturalia, habits of connoisseurship were diffused (if not democratized) in twentieth-century American university art departments. Although Montgomery eschews an exclusively aesthetic focus, the method he devised in relation to furniture, silver, textiles, and ceramics retains a primary emphasis on beauty and workmanship. His heuristic begins: "When first looking at an object, it is important to let oneself go and try to get a sensual reaction to it. I ask myself: Do I enjoy it? Does it automatically ring true? Does it sing to me?" Charles F. Montgomery and Walpole Society (US), *Some Remarks on the Practice and Science of Connoisseurship* (New York, 1961).

43. Klippart, *Wheat Plant*, 30.

44. Daniel J. Browne, "The Production of New Varieties of Wheat by Cross-Fecundation," NARA RG 16.1, XIV.177; printed in *Patent Office Report, Agriculture*, 1856, 247.

45. Klippart, *Wheat Plant*, 35, 81.

46. Cronon, *Nature's Metropolis*, 116–36; Postel, *Populist Vision*, 116 ff.

47. Koselleck, *Futures Past*.

48. Connolly, *A World of Becoming*. Connolly suggests a political theology of immanence or becoming as a way of belonging in time.

EPILOGUE

1. On the Syrian Fertile Crescent, Antun Sa'adah, *The Genesis of Nations* (Beirut: Department of Culture of the Syrian Social Nationalist Party, 2004); P. J. Vatikiotis, "The Politics of the Fertile Crescent," in *Political Dynamics in the Middle East*, ed. Paul Y. Hammond and Sidney S. Alexander (New York: American Elsevier, 1972); Majid Khadduri, "The Scheme of Fertile Crescent Unity," in *The Near East and the Great Powers: With an Introd. by Ralph Bunche*, ed. Richard Nelson Frye (Port Washington, NY: Kennikat Press, 1969); James C. Drewry, "An Analysis of the 1949 Coups D'etat in Syria in the Light of Fertile Crescent Unity" (master's thesis, American University of Beirut, 1960); Fred Haley Lawson, *Constructing International Relations in the Arab World* (Stanford, CA: Stanford University Press, 2006), 41–46; Bonnie F. Saunders, *The United States and Arab Nationalism: The Syrian Case, 1953–1960* (Westport, CT: Praeger, 1996), 43.

2. A valuable case study of this grading of knowledge is the controversy over the detection of transgenic maize contamination in Mexican fields, which elevated the proofs of genetic sequencing technologies above the judgments of farmers, agronomists, and biologists as to the presence of transgenic maize. Christophe Bonneuil, Jean Foyer, and Brian Wynne, "Genetic Fallout in Bio-Cultural Landscapes: Molecular Imperialism and the Cultural Politics of (not) Seeing Transgenes in Mexico," *Social Studies of Science*, October 15, 2014.

3. Rachel Brenchley et al., "Analysis of the Bread Wheat Genome Using Whole-Genome Shotgun Sequencing," *Nature* 491, no. 7426 (2012): 705–10.

4. Ibid.; "Science/AAAS | Special Issue: Slicing the Wheat Genome," July 18, 2014, http://www.sciencemag.org/site/extra/wheatgenome/; Klaus F. X. Mayer et al., "A Chromosome-Based Draft Sequence of the Hexaploid Bread Wheat (*Triticum aestivum*) Genome," *Science* 345, no. 6194 (2014): 1251788.

5. In the last decade historians have turned their attention to how biological sciences can inform historical analysis and characterize relationships between humans and a variety of other species and ecosystems. For example, "Roundtable: History Meets Biology," *American Historical Review* 119, no. 5 (2014): 1492–99; Monica Green has applied these insights to the intersection of molecular genetics and disease history: "Genetics as a Historicist Discipline," *Perspectives on History*, December 2014, http://www.historians.org/publications-and-directories/perspectives-on-history/december-2014/genetics-as-a-historicist-discipline. My emphasis here is on the historical thinking of practitioners of functional genomics rather than the utility of genetic evidence for new kinds of historical inquiry.

6. Brenchley et al., "Analysis of the Bread Wheat Genome Using Whole-Genome Shotgun Sequencing."

7. Brett Frederick Carver, "Wheat Evolution, Domestication, and Improvement," in his *Wheat: Science and Trade* (Ames, IA: Wiley-Blackwell, 2009), 22–23.

8. Some have questioned these distinctions, positing a more fluid relation between farmer improved varieties and laboratory products. E.g., Berg, "Landraces and Folk Varieties."

9. David Vaver, *Intellectual Property Rights: Critical Concepts in Law*, vol. 1 (London: Routledge, 2006), 266, documenting personal communication with Celia Sperling, USDA, 1992.

10. Jonathan W. Silvertown and Amy Whitesides, *An Orchard Invisible: A Natural History of Seeds* (Chicago: University of Chicago Press, 2009).

11. Koselleck, *Futures Past: On the Semantics of Historical Time.*

12. Jason W. Moore, *Capitalism in the Web of Life: Ecology and the Accumulation of Capital* (New York: Verso, 2015).

13. Roy Scranton, *Learning to Die in the Anthropocene: Reflections on the End of a Civilization* (San Francisco: City Lights Books, 2015).

14. Anna Lowenhaupt Tsing, *The Mushroom at the End of the World: On the Possibility of Life in Capitalist Ruins* (Princeton, NJ: Princeton University Press, 2015).

15. Natasha Myers, "Photosynthesis," in Cymene Howe and Anand Pandian, "Lexicon for an Anthropocene Yet Unseen," Theorizing the Contemporary, Cultural Anthropology Website, January 21, 2016, http://www.culanth.org/fieldsights/790-photosynthesis; "Conversations on Plant Sensing: Notes from the Field," *NatureCulture* 3 (2015): 35–66.

16. Katharine Berry Judson, *Myths and Legends of the Great Plains* (Chicago: A. C. McClurg, 1913), 19–20.

INDEX